Wireless RF Energy Transfer in the Massive IoT Era

Wireless RF Energy Transfer in the Massive IoT Era

Towards Sustainable Zero-energy Networks

Onel Alcaraz López
University of Oulu

Hirley Alves
University of Oulu

Library of Congress Cataloging-in-Publication Data

Names: López, Onel Alcaraz author. | Alves, Hirley, author.
Title: Wireless RF energy transfer in the massive IoT era : towards sustainable zero-energy networks / Onel Alcaraz López, University of Oulu, Hirley Alves, University of Oulu.
Description: Hoboken, New Jersey : John Wiley & Sons, [2022]
Identifiers: LCCN 2021033393 (print) | LCCN 2021033394 (ebook) | ISBN 9781119718666 (hardback) | ISBN 9781119718680 (pdf) | ISBN 9781119718697 (epub) | ISBN 9781119718703 (ebook)
Subjects: LCSH: Wireless power transmission. | Internet of things–Power supply.
Classification: LCC TK3088 .L67 2021 (print) | LCC TK3088 (ebook) | DDC 621.319–dc23
LC record available at https://lccn.loc.gov/2021033393
LC ebook record available at https://lccn.loc.gov/2021033394

Cover image: © Shine Nucha/Shutterstock
Cover design by Wiley

Contents

Preface

Why We Wrote This Book

Even though there is vast literature on radio-frequency energy harvesting and wireless energy transfer, the majority of the works do not focus on Internet of Things network deployments and their inherent characteristics and stringent requirements, particularly, the need to support the operation of a massive number of low-power low-cost devices. Powering this massive number of devices is challenging and cannot be done by adapting conventional models designed for a limited number of users. Massive Internet of Things networks will connect thousands to millions of devices in a single cell.

The lack of dedicated solutions prompted us to research in this area, which we call "massive wireless energy transfer" (massive WET). We then decided to gather the most recent contributions on massive radio frequency energy harvesting and compile the information into a single reference that can serve as a guide for future research in this revitalized area. This book reviews the state-of-the-art in radio frequency energy harvesting and points out future research directions. A basic background in statistics and wireless communications is sufficient to read this book, while some basic knowledge of linear algebra, convex optimization, and statistical signal processing may be needed to thoroughly follow the main theoretical derivations and analysis.

Structure of the Book

The book is divided into eight chapters. In a nutshell, Chapter 1 and Chapter 2 introduce the key concepts and provide an overview of the energy harvesting technologies in the context of massive networks. Then, Chapters 3 to 7 delve into analytical and algorithmic aspects of the different protocols, techniques, and scenarios examined in each chapter. Finally, Chapter 8 summarizes the key contributions of each chapter and indicates future research directions.

In Chapter 1, we overview the concept of massive IoT and related connectivity components. Chapter 2 is an overview of the energy harvesting technology based on radio frequency energy transfer for powering future low-power massive network deployments. In Chapter 3, we focus on ambient radio frequency energy harvesting techniques, which exploit readily available environmental radio frequency energy, and review key protocols and scenarios. Then, in Chapter 4, we delve into massive wireless energy transfer, in which energy transmissions from dedicated energy sources energize a massive number of energy harvesting devices, and propose novel energy transfer mechanisms tailored for massive Internet of Things networks. In Chapter 5, we concentrate on a network of power beacons, which are dedicated radio frequency energy transmitter devices. Therein, we discuss and propose optimization approaches for the deployment of a network of power beacons. Next, Chapter 6 discusses the wireless-powered communication network architecture, key characteristics, protocols, and single and multi-user scenarios with respective power control strategies. Chapter 7 overviews the simultaneous wireless information and power transfer architecture, the main receivers, and representative scenarios. Finally, Chapter 8 summarizes the key contributions of this book and indicates future research directions.

How to Use This Book

Researchers who want to delve into the field of radio frequency energy harvesting can read this book from cover to cover. Chapter 1 provides the motivation and introduction to general ideas discussed in the book. Chapter 2 is instrumental to understanding the underlying analytical models used in the remaining chapters. Therefore, after reading Chapter 2, Chapters 3, 4, 6 and 7 can be read in any order, based on personal preference. Even though Chapter 5 can be read independently, it relies on concepts introduced in Chapter 4.

Each chapter ends with a summary of the key points and contributions. A researcher familiar with the broad field of radio frequency energy harvesting can read the summary and become acquainted with the content, and then decide to read in detail.

Onel Alcaraz López and Hirley Alves
Oulu, May 2021

Acknowledgments

We are thankful to our colleagues, collaborators, friends, and family for their constant support and encouraging words throughout the writing of this book.

We thank Richard Demo Souza and Samuel Montejo Sánchez for their detailed comments and insightful ideas. These discussions kept us pushing toward further exploration of our research both in breadth and depth.

We thank Pedro H. J. Nardelli, Dick Carrillo Melgarejo, Nelson J. Mayedo Rodríguez, Dian Echevarría Pérez, Osmel M. Rosabal, Jean de Souza Sant'Ana, Richard Demo Souza and Samuel Montejo Sánchez for their valuable feedback on specific chapters.

We thank Sarah Lemore (Associate Managing Editor, Wiley) for the assistance provided and for putting up with our requests for extensions, which were granted and all were duly appreciated and well used.

We thank our families for their patience, support, and occasionally taking the blame for our lack of skills in time management. Even then, they were always there for us.

O. A. L. & H. A.

Acronyms

3GPP	third generation partnership project
5G	fifth generation of cellular systems
LoRaWAN	LoRa wide area network
LoRa	long range
AI	artificial intelligence
AA–IS	all antennas transmitting independent signals
AA–SS	all antennas transmitting the same signal
ACK	acknowledgment
AF	amplify and forward
AP	access point
AWGN	additive white Gaussian noise
BM	butler matrix
BS	base station
BTB	base transmit band
CCDF	complementary CDF
CDF	cumulative density function
cMTC	critical MTC
CR	cognitive radio
CSCG	circularly symmetric complex Gaussian
CSI	channel state information
D-BPSK	differential binary phase shift keying
D2D	device-to-device
DAS	distributed antenna system
DC	direct current
DF	decode and forward
DLT	distributed ledger technology
DTV	digital TV
EB	energy beamforming
EC-GSM-IoT	extended coverage GSM IoT
eDRX	extended discontinuous reception
EE	energy efficiency

EH	energy harvesting
EMF	electromagnetic field
FD	full-duplex
FDMA	frequency-division multiple access
FEIPC	fixed-error inversion power control
GA	genetic algorithm
GIM	greedy iterative movement -based PBs deployment
GPP	Ginibre point process
GSM	global system for mobile communications
HAP	hybrid AP
HARQ	hybrid automatic repeat-request
HPB	hybrid PB
HPPP	homogeneous PPP
HSU	harvest-store-use
HTT	harvest-then-transmit
HU	harvest-use
HUS	harvest-use-store
HVAC	heating, ventilation, and air conditioning
i.i.d.	independently and identically distributed
IC	integrated circuit
ICL	Imperial College London
ICT	information and communications technology
ID	information decoding
IoE	Internet of Energy
IoT	Internet of Things
IPM	interior point method
IRS	intelligent reflecting surface
ISM	industrial, scientific and medical
ITU	International Telecommunication Union
KPI	key performance indicator
KWW	Kohlrausch-Williams-Watts
LCEB	low-complexity energy beamforming
LIS	large intelligent surface
LOS	line-of-sight
LP	linear program
LPWA	low power wide-area
LPWAN	LPWA network
LTE	long-term evolution
LTE-M	LTE for machines
MAC	medium access control
MCL	maximum coupling loss
MEMS	micro-electromechanical systems
MIMO	multiple-input multiple-output

MISO	multiple-input single-output
mMIMO	massive MIMO
MMSE	minimum mean square error
mMTC	massive MTC
MPE	maximum permissible exposure
MRC	maximum ratio combining
MRT	maximum ratio transmission
MTB	mobile transmit band
MTC	machine-type communications
MTD	machine-type device
NB-IoT	narrowband IoT
NEM	network energy minimization
NLOS	non-LOS
NOMA	non-orthogonal multiple access
OFDM	orthogonal frequency division multiplexing
OFDMA	orthogonal frequency division multiple access
PB	power beacon
PCE	power conversion efficiency
PDF	probability density function
PGFL	probability generating functional
PMU	power management unit
PPP	Poisson point process
PS	power splitting
PSM	power saving mode
PSO	particle swarm optimization
PU	primary user
QCQP	quadratically constrained quadratic program
QoS	quality-of-service
RAB	rotary antenna beamforming
RAT	radio access technology
RF	radio frequency
RPS-EMW	random phase sweeping with energy modulation/waveform
SA	switching antennas
SAR	specific absorption rate
SBS	small-cell BS
SDG	sustainable development goals
SDMA	spatial-division multiple access
SDP	semi-definite programm
SDR	semi-definite relaxation
SI	self interference
SIC	successive interference cancellation
SINR	signal-to-interference-plus-noise ratio
SIR	signal-to-interference ratio

SISO	single-input single-output
SNR	signal-to-noise ratio
SS	spatial switching
SVD	singular-value decomposition
SWIPT	simultaneous wireless information and power transfer
TDMA	time-division multiple access
TS	time switching
UAPB	unmanned aerial PB
UAV	unmanned aerial vehicle
UE	user equipment
ULA	uniform linear array
UN	United Nations
URLLC	ultra-reliable low-latency communications
WET	wireless energy transfer
WIT	wireless information transfer
WPCN	wireless-powered communication network
WSN	wireless sensor network
WUR	wake-up radio
ZF	zero forcing

Mathematical Notation

$\mathbb{I}(A)$	*indicator function. It is equal to 1(0) if A is true(false)*
$[x]^+$	$\max(x, 0)$
$[x]^\times$	$\min([x]^+, ref)$, *where ref refers to certain reference level according to the application scenario, e.g., $B_{\max}$*
$\sim$	*distributed as*
$\mathbb{N}$	*set of natural numbers*
$\mathbb{R}^d$	*real coordinate space of dimension $d \in \mathbb{N}$. d is omitted when equals to 1*
$\mathbb{R}^+$	*set of positive real numbers*
$\mathbb{C}^d$	*complex coordinate space of dimension d. d is omitted when equals to 1*
$\|\cdot\|$	*denotes cardinality or absolute value operation in case of a set or a scalar as input, respectively*
$\|\cdot\|_p$	ℓ_p *norm. The value of p may or may not be specified when $p = 2$*
$\mathbb{E}_X[\cdot]$	*expected value with respect to random variable X. When X is not specified, the expected value is with respect to all random variables*
$\mathbb{P}[A]$	*probability of occurrence of event A*
$\mathcal{L}_x(s)$	*Laplace transform of random variable X*
$\mathcal{L}(\cdot)$	*inverse Laplace transform operator*
$f_X(x)$	*PDF of random variable X*
$F_X(x)$	*CDF of random variable X*
$\tilde{F}_X(x)$	*CCDF of random variable X*
$\Gamma(\cdot)$	*gamma function*
$\Gamma(\cdot, \cdots)$	*upper incomplete gamma function*
$K_\nu(\cdot)$	*modified Bessel function of the second kind and order ν*
$\mathrm{erf}(\cdot)$	*error function*
$\mathrm{Tr}(\cdot)$	*matrix trace operator*
$\mathcal{O}(\cdot)$	*big-O notation*
$\mathrm{diag}(\mathbf{x})$	*diagonal matrix with the main diagonal from entries of $\mathbf{x}$*

$(\cdot)^T$	*transpose operator*
$(\cdot)^H$	*Hermitian transpose operator*
$\text{rank}(\cdot)$	*rank operator*
$\succeq$	*generalized greater-than-or-equal-to inequality: between vectors, it represents component-wise inequality; between symmetric matrices, it represents matrix inequality*
$\inf \cdot$	*infimum operator*
i	*imaginary unit, i.e., $\sqrt{-1}$*
$\mathbb{V}_X[\cdot]$	*variance with respect to random variable X. When X is not specified, the variance is with respect to all random variables*
$J_n(\cdot)$	*Bessel function of first kind and order n*
$\det(\cdot)$	*determinant operation*
$\text{mod}(a, b)$	*a modulo b operation*
$\text{atan2}(c)$	*returns the angle in the Euclidean plane between the positive x axis and the ray to the point c*
$\frac{df(x)}{dx}$	*derivative of single-variable function f*
$\frac{\partial f(\cdot)}{\partial x}$	*partial derivative of multi-variable function f with respect to x*
$\nabla^2 f$	*Hessian of function f*
$Q(\cdot)$	*Q-function, the tail distribution function of the standard normal distribution*

Distributions

$\text{Exp}(x)$	*exponential random variable with mean x*
$\mathcal{N}(\mathbf{m}, \mathbf{R})$	*Gaussian real random vector with mean vector $\mathbf{m}$ and covariance matrix $\mathbf{R}$*
$\mathcal{CN}(\mathbf{m}, \mathbf{R})$	*circularly-symmetric complex Gaussian random vector with mean vector $\mathbf{m}$ and covariance matrix $\mathbf{R}$*
$\chi^2(m, n)$	*non-central chi-squared distribution with m degrees of freedom and parameter n*
$\Gamma(m, n)$	*Gamma distribution with shape m and scale n*

About the Companion Website

This book is accompanied by companion website includes a number of resources created by the author that you will find helpful.

www.wiley.com/go/Alves/WirelessEnergyTransfer

This website includes

- PowerPoint Slides
- PDF

1

Massive IoT

The Internet of Things (IoT) revolution took shape during the past decade, though its roots trace back to the 50s when Alan Turing proposed machines that could sense and learn from their environment. Since then, many pundits have shared their views around this idea. It is likely that the term "Internet of Things" was coined in 1999 by Kevin Asthon; but many years later the term was added to, for instance, Oxford English dictionary [1].

Nowadays, the IoT is much more than a technological revolution because it shapes the surrounding technologies and influences society and industry.

The IoT influences our daily lives, at home and work, the way we do business, and how businesses operate. Some even project that it will change the whole value chain of many businesses in several industrial verticals [2–4], thus becoming a vital component of the 4th industrial revolution, known as Industry 4.0 [5, 6].

The IoT can be defined as the myriad of interconnected objects that can gather, process, and communicate information. Thus, the IoT is a network of physical devices, capable of sensing and gathering information from their environment, and then able to exchange and gather information from other devices or from the Internet.

Therefore, for this book, any object called IoT device, outfitted with proper sensors and actuators, can interact with its surroundings, and with the right communication technology, can share information.

Many of the so-called smart objects available in the market today are IoT devices. For instance, a smart refrigerator that can tell the expiring date of the groceries, then predict the grocery list and inform its owner via a smartphone application; thermostats that sense the environment and regulate the temperature autonomously or under smartphone application; LED lights that regulate colors and brightness according to the time of day, ambient luminosity, or through commands; animal identification tags that broadcast animals' location, vital signs to a designed application.

Wireless RF Energy Transfer in the Massive IoT Era: Towards Sustainable Zero-energy Networks, First Edition. Onel Alcaraz López and Hirley Alves.
© 2022 by The Institute of Electrical and Electronics Engineers, Inc. Published 2022 by John Wiley & Sons, Inc. Companion website: www.wiley.com/go/Alves/WirelessEnergyTransfer

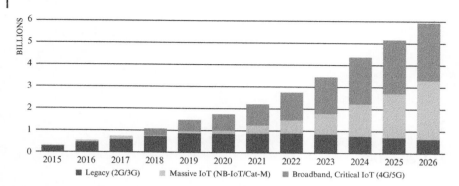

Figure 1.1 IoT connectivity growth trends (2015-2026). *Source:* Data from [8].

Frost & Sullivan report in [7] that the IoT is fundamental to the future of the consumer industry in general by enabling services that improve users' experience. The report identifies the main IoT growth drivers, which are in the domains of 1. data generation from everyday products, 2. different verticals (industries and markets), and 3. ecosystem and advanced technologies integration; while acknowledging the increased need for data privacy and security as the IoT becomes more pervasive. Frost & Sullivan projections correlate with Gartner's recent estimation of 5.8 billion installed IoT devices in 2020 spread into different market segments [2], ranging from automotive, agriculture, retail, to smart buildings and homes.

In this line, Figure 1.1 helps to exemplify the IoT growth trends. The figure shows an estimation on the number of IoT connections from 2015 to 2026 based on cellular IoT connectivity solutions with data provided by Ericsson [8]. Notice that massive IoT solutions will experience a three-fold increase in the next five years while taking market share from the legacy network (namely 2G and 3G).

Although ideas around the concept of the IoT are accumulating for decades, it was not until the last decade that the number of IoT devices and applications started to grow exponentially[1].

A few factors contribute to its popularity in the recent years, for instance: 1. advances in distributed cloud-based computing enabling device data and agile management; 2. advances, even though on a smaller scale, in energy storage and harvesting technologies enabling more extensive lifetime and flexible deployment; 3. overall cost reduction of electronic components (for instance, sensors and actuators).

---

1 See examples at https://www.ericsson.com/en/mobility-report/mobility-visualizer.

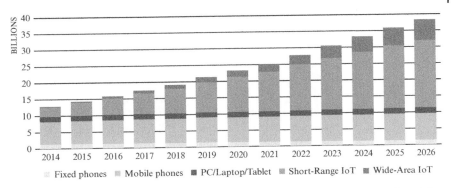

Figure 1.2 Estimate of the number of connected devices until 2026. *Source:* Data from [8].

Nonetheless, one of the key reasons for the fast-paced growth is due to the introduction of novel connectivity solutions for massive deployment of devices, via the so-called massive machine-type communications (mMTC) [3, 4].

Massive IoT comprises applications with a large number (ranging from dozens to millions to billions) of devices, so-called machine-type devices (MTDs). These devices transmit small data packets and are low cost and low energy, and possess heterogeneous and sporadic traffic patterns.

In the context of mobile communications, massive IoT and mMTC share such inherent characteristics.

Moreover, as the name implies, the key characteristic of massive IoT is primarily scalability and versatility. Despite the small information payload that the messages carry, the network must be able to connect such large number of MTDs and offer enough capacity, while providing features that extend battery life and coverage. Because, in many cases, these devices may face challenging radio conditions and may rely only on their battery supply.

For example, Business Insider predicts more than 64 billion MTDs by 2025, a six-fold increase in less than a decade [9]. Ericsson predicts a seven-fold increase in the number of massive cellular IoT worldwide from 2021 to 2026, reaching more than 2.5 billion deployed devices [8].

Figure 1.2 shows a projection made by Ericsson [8] on the number of connected devices until 2026. We can observe that the number of (wide-area) IoT connections based on cellular IoT technologies (discussed in Section 1.2) has more than doubled in the last five years while forecasting a three-fold increase in the next five years. This projection indicates the increasing number of applications and services in different industry verticals and the massive connections that future networks shall handle. To this end, there exist a wide range of technologies that address the challenges of massive connectivity. Even though scalability (massiveness) is the most recognized feature of the massive IoT, low device cost, extended coverage, and long battery life are also often associated with massive IoT.

Because of common characteristics, these technologies fall under the umbrella of the so-called low-power wide-area networks (LPWANs), which are characterized by [10] 1. scalability - support a massive number of connections; 2. long-range - up to tens of kilometers; 3. low power - battery-powered devices with 20 years lifetime; and 4. low cost - use of license-free bands, simple hardware design and lightweight protocols associated with star or mesh topologies reduce infrastructure and device cost.

In the next section, we review the main LPWA technologies and their evolution towards the next generation of mobile communications.

Our objective is to point out a few predominant technologies in specific industry verticals, discuss their pros and cons, and argue for harmonious and convergent coexistence in future wireless networks, while laying the path for the following chapters that delve into wireless energy transfer (WET) techniques, that even though general, have a mobile communication network as a backdrop. Moreover, we delve into massive IoT, we introduce key definitions, scenarios, and performance and value indicators, and we discuss about promising technologies and enablers. We close this chapter discussing the power problem that challenges massive IoT deployments, whose solutions we develop in the following chapters.

1.1 Selected Use-cases and Scenarios

There is a large variety of use-cases in many industries and societies. Herein we highlight a few, while in Chapter 2, we provide even more examples in the context of massive WET.

The number of potential applications is as massive as number of devices in massive IoT, with diverse requirements in cost, energy efficiency, coverage, performance (for instance, reliability, latency, throughput, capacity, to name a few), and security.

Many IoT devices, namely MTDs, send only a few messages on a given period, such as an hour, or a day or week, while others may send more frequent updates or need higher data rate, voice traffic, or security. Figure 1.3 provides some use-cases and applications of massive IoT.

Smart metering, such as for electricity, water, or gas, is a successful example of massive IoT deployment. In this case, devices often require periodic updates on consumption readings to a remote server or cloud. Moreover, the smart meter can operate under alarm mode when an outage happens, and receive requests or commands remotely.

E-health life sign monitoring and wearables may require a minute or even second-level updates, which demand reliable connections and increase traffic over the network. In large medical facilities, positioning and tracking can be

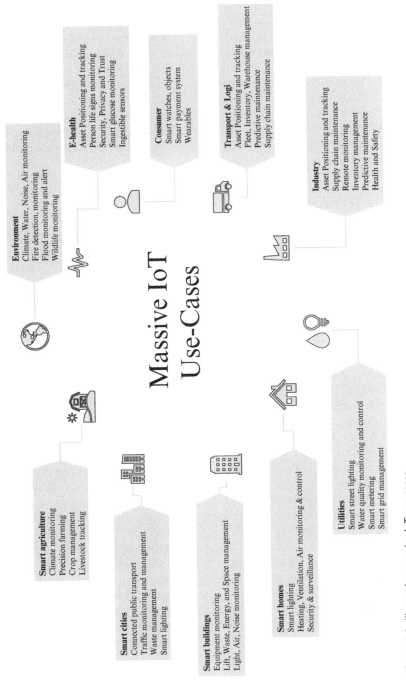

Environment
Climate, Water, Noise, Air monitoring
Fire detection, monitoring
Flood monitoring and alert
Wildlife monitoring

E-health
Asset Positioning and tracking
Person life signs monitoring
Security, Privacy and Trust
Smart glucose monitoring
Ingestible sensors

Consumer
Smart watches, objects
Smart payment system
Wearables

Transport & Logi
Asset Positioning and tracking
Fleet, Inventory, Warehouse management
Predictive maintenance
Supply chain maintenance

Industry
Asset Positioning and tracking
Supply chain maintenance
Remote monitoring
Inventory management
Predictive maintenance
Health and Safety

Massive IoT Use-Cases

Smart agriculture
Climate monitoring
Precision farming
Crop management
Livestock tracking

Smart cities
Connected public transport
Traffic monitoring and management
Waste management
Smart lighting

Smart buildings
Equipment monitoring
Lift, Waste, Energy, and Space management
Light, Air, Noise monitoring

Smart homes
Smart lighting
Heating, Ventilation, Air monitoring & control
Security & surveillance

Utilities
Smart street lighting
Water quality monitoring and control
Smart metering
Smart grid management

Figure 1.3 Illustrative massive IoT use-cases.

realized in real-time, facilitating resource management and improving efficient care. These cases need deep indoor coverage as the extensive facilities have many floors and underground corridors connecting different places.

Other applications in agriculture, transport, and industry may require voice commands and higher data rates. For instance, in smart agriculture, monitoring climate conditions and crop management is essential for precision farming since sensors gather information on climate conditions and humidity, leaf water potential, and other characteristics to monitor crop health. This data is collected and conveyed periodically, for example, hourly or daily. Some applications frequently upload images to a cloud (or remote server), which demands higher throughput.

1.2 Key Technologies

LOng RAnge (LoRa). LoRa is a proprietary long-range, low-power communication technology that operates at sub-GHz unlicensed frequency bands, namely industrial, scientific, and medical (ISM) frequency bands. LoRa employs chirp spread spectrum and uses six practically orthogonal operational bands, denoted as spreading factors, which allow rate and range adaptation [11].

LoRa wide area network (LoRaWAN), regulated by LoRa Alliance [12], uses LoRa as its physical layer. LoRaWAN builds upon a star of stars topology where devices communicate in a single-hop with a gateway, which connects to a network server via standard IP protocol. The medium access control (MAC) is ALOHA-like. LoRa characteristics enable multiple devices to communicate simultaneously using different spreading factors, thus supporting 65536 connections per gateway [11].

LoRaWAN provides several configurations based on six distinct spreading factors and bandwidths, usually 125 kHz or 250 kHz for uplink, depending on regional regulation. The distinct spreading factor configuration allows symbol duration and transmission rate flexibility. For instance, higher spreading factors extend the symbol duration, thus increasing robustness with a lower rate. Such flexibility comes at the cost of increased time-on-air, yielding channel usage increase, thus leading to higher collision probability as extensively discussed in [13]. However, code replication [14], superposition [15], adaptive data rate [16, 17] are techniques that help mitigate these effects. Since the technology operates in the ISM band, it is vulnerable to interference from neighboring deployments and other technologies impacting the performance [18].

Unlike other proprietary solutions, LoRa is open, with a relatively vast amount of information about its operation, thus being one reason for its popularity in many communities, particularly the academic community. Many research works in a broad range of topics (propagation measurements, network performance, and characterization, simulators, among others) in recent years. See, for instance, [13–20] for a comprehensive survey of the recent advances.

SigFox. This technology operates in a star topology, and devices connect to any base station (BS) in range using ultra narrowband signals of 100 Hz in ISM frequency bands. The signal carries a 12-byte payload and travels two seconds over the air, with a fixed data rate of 100 or 600 bits/s using D-BPSK modulation. These characteristics enable reduced energy consumption and broad coverage [21]. Unlike LoRa, much less information is available about SigFox. Only a few works are analyzing the technology, for instance [22–24].

Long Term Evolution for Machines (LTE-M). The mobile communications industry has an ever-growing interest in the IoT and has coined the term MTC leading to new services.

MTC operation mode was introduced in Release 12 (namely LTE category 0 (Cat 0)) to support communication via carrier network in a cost-effective and power efficient way compared to the legacy network. Thus, LTE Cat 0 constitutes a preliminary attempt to somewhat support IoT connectivity, and comprises a low complexity machine-oriented specification but with standard control information under full carrier bandwidth. Enhanced MTC, introduced in Release 13 (Cat M1), mitigates this issue, reducing spectral and energy wastage.

Later releases continued the enhancements and that led to improvements in coverage, latency, battery life, and capacity[2] with the changes in bandwidths (from 1.4 MHz in Cat 0 to 5 MHz in the evolution), and half-duplex operation, and features such as power saving mode (PSM) and extended discontinuous reception (eDRX) [9, 25].

Besides global coverage, LTE-M supports high data rates, real-time traffic, mobility and voice, which in turn makes this technology versatile, unlike other technologies discussed so far, although at cost of complex architecture and costly transceiver.

Narrowband-IoT (NB-IoT). NB-IoT inherits many of its features from LTE, and as its name suggests, it focuses on narrowband signals that occupy 180 kHz bandwidth, which corresponds to one resource block in the legacy LTE system. NB-IoT design is leaner than its counterpart, LTE-M, to match the requirements of many battery-constrained IoT applications. Therefore, it offers good indoor coverage, support to massive connectivity, low power consumption and optimized network architectures (also a characteristic in LTE-M), while

---

2 The interested reader may refer to [9, 25–27] for a thorough discussion on LTE-M evolution, which is out of the scope of this book.

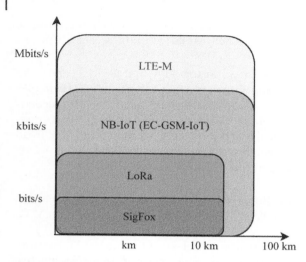

Figure 1.4 Comparison of LPWA technologies in terms of cell coverage and data rate. *Source:* Illustrative numbers based on [29].

operating under licensed spectrum [9, 28]. These two cellular technologies differ vastly by their target.

Both, NB-IoT and LTE-M, are optimized to reduce power consumption, extending devices battery life, though NB-IoT has even lower device power consumption and lower chipset cost. One advantage of these technologies is that both support in-band deployment with the legacy LTE (including LTE guard-bands in NB-IoT), or as standalone with dedicated spectrum.

A recent comparison of these LPWA technologies is available in [9, 27, 29][3]. Figure 1.4 compares the LPWA technologies with respect to cell coverage and data rates. As discussed above, unlicensed technologies have limited data rates and cell coverage, usually ranging from few kilometers in urban environments up to tens of kilometers in rural areas, when compared to licensed technologies. On the other hand, device cost and network deployment are often advantageous. However, cellular technologies provide quality-of-service (QoS) guarantees via long-term service level agreements. At the same time, operation under licensed spectrum comes with the advantage of predictable and controllable interference through efficient mechanisms that enable a massive number of connections.

Massive MTC (mMTC). Massive IoT flourishing markets impose increasingly challenging connectivity demands. Third generation partnership project (3GPP) has taken revolutionary steps in Release 12 when introducing MTC,

---

3 One more cellular IoT standard is the extended coverage GSM IoT (EC-GSM-IoT), which is optimized to harness the already available GSM infrastructure deployed in previous generations, thus operating in-band with GSM [30].

and evolutionary ones in posterior releases by enhancing the performance of LTE-M and introducing new services (NB-IoT and EC-GSM-IoT introduced in Release 13).

MTC revolutionized the mobile communications industry by shifting the focus from broadband services towards the IoT. Clear evidence comes with the fifth generation (5G) that inherently dedicates two-thirds of the service modes to the IoT. In 5G, MTC branched out to massive MTC (mMTC) and critical MTC (cMTC), also known as ultra-reliable low-latency communications (URLLC)[4].

mMTC stands for massive referring to large number of users connected to the network, widely expected in many IoT applications. Therefore, in the communications community, mMTC and massive IoT are used as synonyms and interchangeably.

Even though mMTC is encrusted 5G jargon, it is an umbrella term that specifies the ensemble of solutions toward massive IoT. Thus, it comprises the cellular IoT technologies discussed so far, namely LTE-M and NB-IoT, and solutions under development toward current and next generations, as well as non-cellular technologies, such as LoRa and SigFox. This is because 1. LTE-M and NB-IoT are compliant with the evolution of mobile communications, thus both solutions operate in-band with 5G, and will evolve (so-called future proof) complying with 5G mMTC requirements [9, 25]; 2. mMTC encloses LPWANs; therefore, non-cellular LPWA technologies fall within this definition. Another point is that even with the advantages of cellular IoT, it is unlikely that a single technology will be ubiquitous in a fragmented market. Therefore, future generations beyond 5G are likely to coexist and complement unlicensed solutions as foreseen in [33].

mMTC key challenges are:

- Energy efficiency
  In many applications, devices rely solely on batteries, and very often, replacement is costly, dangerous, or simply not possible (see Chapter 2). Though battery lifetime may be extended with smart sleeping mode techniques, it may not be sufficient.
- Scalability
  Support to a massive number of connections. The network capacity should also scale to accommodate the demands of such large number of devices. Figure 1.1 illustrates the projected growth for cellular IoT solution.
- Coverage
  Deep indoor coverage is a crucial requirement for many applications, requiring regional, national, or even global coverage.
- Heterogeneity
  Different applications impose different requirements, e.g., in terms of data rates, latency, reliability, energy efficiency, coverage. Flexible connectivity is imperative to handle heterogeneous requirements.

---

4 For further details on URLLC please refer to [31, 32].

- Device cost
  Cost is a critical factor in meeting the economies of scale and many use-cases. For instance, cellular IoT solutions have reduced peak rate and device complexity, and half-duplex operation and narrow bandwidths help address this challenge.

These are the most representative challenges of mMTC, although other capabilities exist, as initially identified by the ITU [34], e.g., peak and experienced data rates, spectrum efficiency, mobility, latency, reliability, and security. In this context, cellular IoT becomes advantageous by coping with the LPWAN requirements and key performance indicators (KPIs) while pushing some of these to their limits, whereas handling security and QoS requirements [4].

1.3 Requirements and KPIs

mMTC challenges arise from related massive IoT challenging requirements. The essential requirements and KPIs for massive IoT are:

- Energy efficiency
  Battery life becomes an aggregated performance indicator for energy efficiency. Current technologies target 10 to 20 years of lifetime.
- Scalability
  Support to a massive number of connections. A target for mMTC is one million MTDs per square kilometer.
- Coverage
  Deep indoor coverage is a crucial requirement for many applications. Overall, the maximum coupling loss (MCL) of 164 dB is the target for current solutions.
- Device cost
  The cost impacts scalability and market penetration. Most of the current chipsets available in the market average around 10 dollars.
- Data rates
  Many applications demand few bytes of information to report, thus a data rate of a few hundreds of bps is sufficient. However, other applications upload photo or video to a remote server, or cloud, which require a few kpbs or even Mbps. Minimal throughput is set to 160 bps in NB-IoT.
- Security
  It comprises security, privacy, authentication, encryption, integrity protection of user data, and denial of service attacks, and the demands vary for each application.

In 2015, the International Telecommunication Union (ITU) identified KPIs for massive connectivity, which has served as a guide for the technology evolution [34].

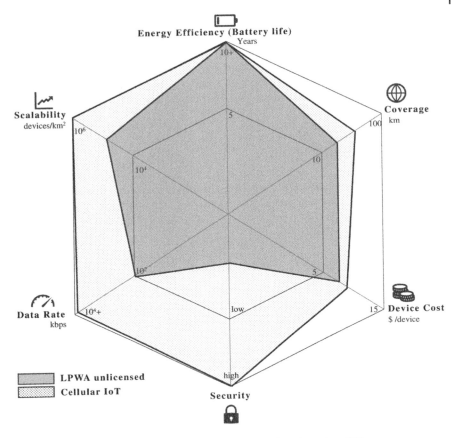

Figure 1.5 KPIs for massive IoT. *Source:* Illustrative numbers based on [29].

The identified KPIs are: peak and user experienced data rate, spectrum efficiency, mobility, latency, connection density, network energy efficiency, and area traffic capacity. Out of these, connection density received the highest importance among ITU ranking, followed by network energy efficiency. Current technologies have targeted these KPIs and offered many solutions to enhance connection density, battery life, and network energy efficiency.

Figure 1.5 captures the most relevant requirements and KPIs for our discussion. For this figure we split the technologies into two major groups: unlicensed and licensed (also known as cellular IoT). The latter comprises technologies based on mobile communications and addresses most of these requirements as illustrated in the figure; while the former comprises solutions build upon ISM frequency bands, which are unlicensed.

All solutions have their pros and cons, as discussed in Section 1.2. Therefore, when choosing between licensed and unlicensed, it is important to pay attention to the application needs and the trade-offs between different technologies. In general, cellular IoT device's cost and service fees tend to be higher than unlicensed alternatives. On the other hand, unlicensed options have limited capabilities concerning scalability, coverage, and security.

1.4 Key Enablers

In this section, we discuss some enablers for efficient connectivity in massive IoT deployments. First, we discuss a holistic and scalable system, which aims to address the networking and the potential of global scalability. Next, we focus on enablers for sustainable connectivity.

1.4.1 Holistic and Globally Scalable Massive IoT

Holistic Massive IoT. With the increasing demands of massive IoT applications, the network should support dynamic and collaborative orchestration between end-to-end applications, often communicating over multiple domains (wireless, wired, optical), involving multiple radio access technologies (RATs) operated by different stakeholders (public, private, cellular). Thus, the network handles the requirements, even though the application becomes agnostic of the interfaces [35]. For instance, [33] proposes a step toward such integration, since it envisions LPWANs complementing cellular services in many massive IoT applications. At the same time, LPWAN resorts to cellular IoT (and broadband) to extend and supplement its services and applications. However, the volume information and network parameters to be optimized grow exponentially, increasing management and orchestration complexity. In this context, solution based on machine learning and artificial intelligence (AI) may be helpful allies in handling this challenge [36, 37].

Global Scalability. One of the most significant advantages of cellular IoT is global availability, allowing fast penetration of massive IoT applications in many markets. However, competing unlicensed technologies can leverage extensive coverage. In special, non-terrestrial networks, such as low-Earth orbit satellites, unmanned aerial vehicle (UAV), high-altitude platforms, have emerged in recent years as strong competitors, particularly for rural, off-shore, and overall remote connectivity [38, 39]. The combination of mMTC and non-terrestrial networks seems to be a vital component of the next generation of cellular systems, aligning with the idea of holistic massive IoT, as discussed in [35]. Some recent works have already shown promising results in this direction; for instance, [40] analyses the performance of mMTC and satellites in a smart city context, while authors in [41] propose a direct connection between terrestrial MTDs and the satellite. Extending this idea, [38] conceptualizes two new architectures to integrate mMTC and satellites, with a direct and an indirect connection between these two entities. These promising findings evince the feasibility of these options even with typical current LPWAN configurations.

1.4.2 Sustainable Connectivity

This section identifies technologies that are the building blocks of sustainable information and communications technology (ICT).

In 2015, the United Nations (UN) defined 17 intertwined sustainable development goals (SDG), while calling on society to meet them by 2030. The ICT and wireless networks industry is one of the most prominent industries to positively contribute towards all established goals, facilitating the creation and access to new services and reduced CO_2 emissions [42]. In this context, sustainable connectivity is fundamental toward increasing efficiency in using resources in future networks, as pointed out by a group of international experts led by the 6G Flagship framework [43]. The UN indicates in [44] that the ICT sector is instrumental in making a significant impact on emissions reduction.

Recently, the European Commission announced an overarching objective towards a green future that is economic growth decoupled from resource use [45, 46]. Furthermore, the European Commission has prioritized digitization [45] and achieving carbon neutrality through the Green Deal [46] as crucial development goals for a sustainable Europe. The European Union is setting ambitious targets for climate change and sustainability for the coming decades, accompanied by many other countries worldwide, as evinced by the 2020 Climate Change Performance Index Results report [47].

In line with this, in 2021, the Finnish Ministry of Transport and Communications announced its climate strategy for the ICT sector [48], which comprises six objectives and measures paving the path to achieve ecologically sustainable digitization. Interestingly, some of those objectives resonate with the ideas in this book, namely: 1. increase of the efficiency of resource use, 2. boost of the devices lifetime, and 3. reduction of energy consumption by the ICT sector.

The ICT sector consumes from 4% to 10% of global electric power, and it is responsible for 3% to 5% of greenhouse gas emissions [48]. Moreover, a recent study indicates that devices and infrastructure are responsible for about 70% of the total energy consumption in the ICT sector [49].

In light of the SDG and climate action, many stakeholders in the ICT sector have taken action. For instance, two of the largest manufacturers, namely Nokia [50, 51] and Ericsson [52], have set clear targets towards sustainable ICT. Moreover, along with the UN' forecast [44] concerning the use of ICT solutions, Ericsson's research [52] recently predicted a 15% reduction on global carbon emissions by 2030 and decrease to less than 2% on ICT's global carbon footprint.

All in all, ICT plays a pivotal role in sustainability and climate action. However, the ICT section is undergoing a sustainable transformation as well, evinced by energy consumption reduction, energy efficient designs, inclusion of renewable sources [44, 51, 52]. However, devices and infrastructure are still critical contributors to the energy footprint of the ICT sector. Therefore, it is essential to identify technologies and methods that provide energy-saving solutions lowering their footprint. In this line, we identity key technology enablers toward sustainable ICT. We briefly summarize some of these fundamental technologies, and then in Chapter 2, we delve into technical details and analysis.

Intelligent Reflective Surface (IRS). The IRS emerges as a promising alternative solution to reconfigure the wireless propagation environment via software-controlled reflective meta-materials. The IRS builds on numerous low-cost, passive, reflective meta-material elements. These elements induce controlled phase and amplitude variations on an incident signal. Hence the IRS enables a programmable and controllable wireless environment. This controllability empowers communication engineers to better combat the impairments of the wireless channel in a completely new form [53–55]. Moreover, the IRS does not require a complete transceiver chain. Besides, assembly is flexible and can cover arbitrarily shaped surfaces. These two characteristics impact directly cost and energy consumption.

The transition towards high-frequency bands, namely millimeter and sub-millimeter-wave bands, is underway in cellular systems. In this context, the IRS is an option to provide low-cost link diversity or strengthen line-of-sight (LOS) channel components.

Nonetheless, to achieve scalable performance and overcome competing technologies, the number of elements in an IRS needs to be large. This approach is unbearable in a fully programmable system, since each element requires an individual (phase and amplitude) control, and channel measurement, thus imposing communication overheads. AI-based solutions can help mitigating such challenges [54]. Another option is based on channel state information (CSI)-free solutions (discussed in Chapter 2) to reduce the signaling and instantaneous estimation of controllable features.

Tiny Machine Learning. Tiny machine learning, known as TinyML, is a flourishing field devoted to extremely low power, always-on, battery operated devices, which is stereotypical of many MTDs. TinyML objective is to integrate machine learning mechanisms into microcontroller-equipped objects. The idea comes from the fact that as the network and technology evolve, information processing and AI become more distributed. Therefore, it moves from the cloud to the edge, and now from edge to devices. Notably, it goes toward extremely low power devices, which can potentially make AI ubiquitous for many massive IoT use-cases. Besides, TinyML opens up opportunities to novel services and applications that do not require centralized (cloud) processing, which alleviates energy consumption and reduces the risk related to security and privacy [56].

For instance, [57] introduces a flexible toolkit that generates energy efficient and lightweight artificial neural networks for a class of microcontrollers. The authors discuss the different optimizations in energy efficiency and computational power. Besides, they test the applicability of their results in a self-sustainable wearable.

Moreover, [56] provides a comprehensive review of the TinyML ecosystem and frameworks for integrating algorithms into microcontrollers, and proposes a multi-RAT architecture suitable for MTDs comparing the performance of different classification algorithms. The results show that these algorithms achieve good accuracy and speed. However, the authors warn about careful design needed due to memory restrictions at the microcontroller. All in all, these recent findings evince a promising research area.

Heterogeneous Access and Resource Management. Conventionally, orthogonal access and radio resource management ensure that users connect to the network and convey information. Nonetheless, with mMTC, the number of users scales to a point beyond the availability of orthogonal resources, which may incur performance losses, extended delays, large energy consumption. In this context, non-orthogonal solutions emerge at both levels, access and resource management. To combat such losses, non-orthogonal multiple access (NOMA) multiplexes users in either of the two most prominent solutions: power [58] or code [59] domain. This strategy can allocate more users through sophisticated successive interference cancellation (SIC) techniques than the available resources, which outweighs the cost of increased complexity at the receiver. A more recent development called rate-splitting multiple access [60] promises even further improvements in spectral and energy efficiencies than other non-orthogonal counterparts [61, 62]. At the same time, [63] discusses the information-theoretical limits of massive access.

Non-orthogonality plays an essential role in other domains, such as resource management. In [64], the authors introduce and analyze the heterogeneous coexistence of different service modes, referred to as non-orthogonal slicing, from an information-theoretical perspective. Later, this idea is expanded or combined with other (non-)orthogonal access solutions [65–68]. These recent works evince the importance of non-orthogonal massive access, primarily when associated with heterogeneous resource management techniques.

Distributed Antennas and Networks. Distributed antenna system (DAS) comprises geographically separated antennas interconnected to a central processing unit, which handles the signal processing. DAS increases diversity in the network [69]. This theoretical concept permeates the wireless communications community since the early 2000s. However, only recently, it is becoming a reality in practice due to many advancements in signal processing, distributed systems, cloud and edge computing [70–73].

In a cell-free DAS[5], known as cell-free massive multiple-input multiple-output (mMIMO), the access points (APs) (equipped with massive number of antennas) connect via backhaul to central processing units. These processing nodes may belong to a centralized or distributed architecture, for instance [73] evaluates pros and cons of each approach. Together these interconnected APs serve the devices under the coverage area [70], thus reducing interference and achieving higher data rates.

A pivotal distinction to previous solutions is scalability and the ability to [72] 1. provide (almost) uniformly high signal-to-noise ratio (SNR) and data rates across the network; 2. manage interference joint processing multiple APs compared to other solutions that face inter-cell interference; 3. increase SNR via coherent transmissions since APs with weaker channels contribute to joint processing.

Hardware imperfections at the user side and backhaul availability may hamper these gains and are yet open challenges. Incorporating backhaul capacity and processing ability at the edge becomes critical as well [73]. Advances in this area may lead to further improvement in network access, processing time, and energy efficiency.

Short Length Codes. Legacy communication systems design and optimization rely on channel capacity, which assumes infinite long block lengths. Thanks to the law of the large numbers, a code averages out the stochastic variations imposed by the wireless channel. The notion of capacity is a reasonable benchmark in many practical systems since packets comprise a long block of several bits (in the order of thousands of bits). This fundamental result arose in the seminal work of Claude Shannon [74].

However, in many massive IoT applications, the information payload of the message is relatively short, only a few bytes. In this case, current state-of-the-art codes perform poorly, which results in wastage of resources and poor performance.

Only recently, new information-theoretic results appeared related to the performance of short-block length communication. These results show that the achievable rate depends not only on the channel quality of the communication link but also on the actual block length error probability tolerable at the receiver [75]. Please refer to Appendix A for an overview of finite block length coding.

---

5 A detailed account of the historical background and foundations of this technique can be found in [72].

Cellular systems rely on low-density parity-check and polar codes, however tuned to serve broadband users. mMTC on the other hand demands many changes on, for instance, 1. the optimization of coding schemes for a short block of length - ranging from a few hundred to a thousand bits; 2. robust error detection in order to avoid the need for outer codes; and 3. novel decoding algorithms able to work with limited or even without CSI at the receiver.

In addition, in the context of WET, the message block length impacts the energy and information reliability, thus raising an interesting trade-off between energy-information reliability and resource allocation [76, 77].

CSI Free Solutions. Traditionally, CSI is used to compensate the impairments imposed by the wireless channel while consequently improving the communication performance. However, CSI acquisition becomes costly in massive IoT due to number of devices, heterogeneous traffic patterns, latency requirements [78]. In light of these constraints, novel CSI-free and CSI-limited (when, for instance, only long term statistics are available) solutions are needed taking also into account environmental conditions and side information to spare energy, time and spectrum resources.

The subsequent chapters carefully discuss promising CSI-free solutions for massive IoT.

Energy Harvesting, Zero-energy Devices and Backscatter. Incorporating energy harvesting (EH) capabilities into the MTDs will significantly impact the next generation of mMTC since MTDs will operate batteryless [78, 79]. Zero-energy devices will require a rethinking of the air interface and network architecture. Pundits forecast more than 40 years battery life and even continuous battery-less operation of zero-energy devices [35, 80].

Whenever the energy transfer happens through radio frequency (RF), the devices can either harness the ambient energy from neighboring transmissions, known as ambient RF–EH (see Chapter 3) or be served by a dedicated carrier, know as WET. The former is discussed within Chapter 2, while the latter is the scope of this book whose objective is to promote the most recent, and in our view, the most promising solutions towards sustainable ICT in massive IoT networks.

1.5 Final Remarks and Discussions

In this chapter, we discussed the exponential growth of the IoT in the recent years, and its potential to impact complete value chains across many sectors, societal and industrial alike. We highlighted how wireless connectivity is instrumental to these changes. Moreover, we introduced and discussed selected use-cases, requirements and KPIs, key technologies and enablers for building sustainable ICT. Before concluding, let us discuss further a fundamental issue in massive IoT, and how it is addressed in this book.

The 'Power Problem' in Massive IoT. The energy use and demands of a MTD may be small, in the order of μW to mW. Nonetheless, powering a large number of devices in the network poses a herculean task. Even though most of the devices are battery-powered and current connectivity technologies promise more than ten years of battery life, a MTD performs much more tasks than communication. Hence, its battery, and therefore the device's lifetime, may run out sooner than anticipated. Thus, the battery needs replacement, and in some cases discarding the whole device. In any case, the generation of electronic waste (e-waste) becomes a serious issue since it posses a health and environmental hazard due to toxic additives or hazardous substances [81, 82].

The UN monitors the generation of e-waste closely. Its recent report [82] indicates a record of more than 53 million tonnes of e-waste in 2020 (21% increase in 5 years), and it predicts more than 74 million tonnes by 2030. Though some initiatives towards recycling are already taking place [83], more actions shall arise in the next decade. Moreover, considering the massive number of devices and shorter life cycle, the volume of person-months for maintenance-related tasks becomes economically unsustainable for many small and larger businesses alike.

All in all, battery-equipped MTDs are not a scalable solution for massive IoT deployments due to frequent replacements that may become dispendious and contribute to the generation of e-waste. In order to tackle this issue, EH (photovoltaic, induction, RF, among others [78, 79]) has become a sound solution for powering such a massive number of devices. A recent white paper from a multidisciplinary expert group indicates WET as a powerful technology for future mMTC [35]. A recent study proposes WET as a critical component toward sustainable ICT.

The RF–EH Solution. In this book, we advocate using RF-based energy transfer and harvesting for massive IoT deployments. We are continuously developing a technology called **massive WET**[6], which from inception targets the power problem in massive deployments of MTDs.

Massive WET techniques [77, 78, 84–88] deployed in a power beacon (PB) allow powering a multitude of devices without imposing communication overhead. Besides, massive WET 1. increases the battery life cycle of MTDs, consequently enhances network lifetime; 2. contributes to the replacement reduction of batteries and devices, thus decreases the generation of e-waste; and 3. facilitates deployment and planning of the network. Due to wireless

---

6 We coined the term massive WET already in 2019 in a pre-print version of [78].

powering, device placement is flexible and even possible in dangerous or difficult to reach places; and 3) incites development of batteryless devices. WET will be ubiquitously available and devices can rely solely on WET to operate.

We envision a ubiquitous wireless power architecture composed of several PBs, employing the massive WET techniques discussed in this book. This architecture coexists and complements the information network (cellular or unlicensed LPWAN) such that a massive number of IoT devices (MTDs, battery-equipped and batteryless alike) profit from guaranteed energy delivery.

2

Wireless RF Energy Transfer: An Overview

2.1   Energy Harvesting

EH technologies play a key role towards enabling sustainable IoT networks. The main reason is that as future IoT networks scale (billions of wirelessly connected MTDs by 2030 [3]), exploiting traditional energy sources relying on batteries become:

- costly since human intervention for maintenance (e.g., battery replacement operations) increases;
- unsustainable since the greater demands for maintenance operation and the battery waste make the carbon footprint soar. In addition, battery waste processing is already a critical problem because of its inherent eco-unfriendly nature;
- unappealing since MTDs' form-factor is in general dominated by the size of traditional batteries, and consequently cannot decrease significantly.

Instead, EH allows a perpetual and flexible operation of MTDs by replacing the power cable and batteries with harvested energy, thus, cutting the very *"last wires"* for truly wireless and autonomous communications [89]. Moreover, properly designed EH-enabled networks are energy-efficient by design, and are consequently of paramount importance towards building a sustainable world as targeted by UN' 2030 SDG[1].

2.1.1   EH Sources

EH technologies for wireless MTDs can be mainly classified into the following two categories:

---

1   See https://www.un.org/sustainabledevelopment/sustainable-development-goals/

Wireless RF Energy Transfer in the Massive IoT Era: Towards Sustainable Zero-energy Networks.
First Edition. Onel Alcaraz López and Hirley Alves.
© 2022 by The Institute of Electrical and Electronics Engineers, Inc. Published 2022 by John Wiley & Sons, Inc. Companion website: www.wiley.com/go/Alves/WirelessEnergyTransfer

Table 2.1 Comparison of the main EH sources.

Cat.	Energy Source	Energy Provision[†]	Hardware Complexity	Main Limitation
Ambient EH	Light (Indoor)	$50\ \mu\mathrm{W/cm^2}$	Moderate	Sensitive to blocking
	Light (Solar)	$50\ \mathrm{mW/cm^2}$	Moderate	Sensitive to blocking
	Thermal	$40\ \mu\mathrm{W/cm^2}$	Moderate	Low efficiency, large area
	RF	$10\ \mathrm{nW/cm^2}$	Low	Low energy density
Dedicated EH	Piezoelectric	$0.3\ \mathrm{mW/cm^3}$	Moderate	Fragile materials
	Electrostatic	$80\ \mu\mathrm{W/cm^3}$	High	Short distance
	Acoustic	$1\ \mathrm{mW/mm^2}$	High	Limited applicability
	Induction	$1\ \mathrm{W/cm^2}$	High	Short distance
	Magnetic resonance	$1\ \mathrm{W/cm^2}$	High	Large form factor
	RF	$10\ \mu\mathrm{W/cm^2}$	Low	Low efficiency, health issues

Source: Based on [90–92].
[†] The energy provision must be interpreted as a measure of order and not as an exact measure. The real values according to specific system implementations may vary significantly around the provided values.

- ambient EH, which relies on energy resources that are readily available in the environment and that can be sensed by EH receivers;
- dedicated EH, which are characterized by on-purpose energy transmissions from dedicated energy sources to EH devices.

Different from dedicated EH-based setups, ambient EH does not require additional resource/power consumption from the surrounding (sometimes newly-deployed) energy network infrastructure. However, temporal, geographical, or environmental circumstances may limit their service guarantee, thus making them inappropriate (at least as a standalone solution) for many use cases with QoS requirements. The main energy sources within the above categories and their associated characteristics are presented in Table 2.1.

The ambient EH methods based on light intensity and thermal energy are highly sensitive to blocking and perform with low conversion efficiency, respectively. But maybe more importantly, both demand an add-on EH material and circuit, which in practice limit the form factor reduction to the desired levels for many use cases. The same strong limitation is characteristic of the induction, magnetic resonance coupling and many piezoelectric-based dedicated EH methods. The induction method is based on the inductive coupling effect of non-radiating electromagnetic fields, including the inductive and capacitive mechanisms, and is subject to coupling misalignment impairments that limit the range and scalability. The magnetic resonance coupling makes use of the well-known principle of resonant coupling, i.e., two resonant objects at the same frequency tend to couple with each other most efficiently. Magnetic resonant coupling is able to achieve higher power transfer efficiency over longer distances than inductive coupling, by carefully tuning the transmitter and receiver circuits to make them resonant at the same frequency. Meanwhile, piezoelectric-based EH relies on the energy coming from a mechanical strain captured by an usually fragile, piezoelectric material layer on top of the wireless device [91]. In general, electrostatic and acoustic methods overcome the devices' size limitation, but they are either limited to very short distance operations or to very specific applications. Specifically, in the electrostatic method, a mechanical motion or vibration is used to change the distance between two electrodes of a capacitor against an electrical field, thus, transforming the vibration or motion into electricity due to the capacitance change [91]. Meanwhile, acoustic WET, which is usually in the range of ultra-sound, is quite efficient but mostly for transferring the energy over non-aerial media such as water, tissue or metals, and are more appropriate for medical applications [90]. Another potential technology for dedicated WET is laser power beaming [93, 94], which uses highly concentrated laser beam aiming at the EH receiver to achieve efficient power delivery over long distances. However, laser radiation could be hazardous, it is highly sensitive to blocking, and requires accurate pointing towards the receiver, which could be challenging to achieve in practice.

In contrast to the EH methods discussed above, the RF-based EH inherently allows

- small-form factor implementation since the same RF circuitry for wireless communications can be re-utilized totally or partially for RF EH;
- native multi-user support since the same RF signals can be harvested simultaneously by several devices.

The above key features, when combined, make RF EH a strong candidate, much more suitable than the EH technologies based on other energy sources, for powering many low-power IoT use cases. When the number of devices increases, RF–EH technologies become even more appealing, and are hereinafter the focus of this book. In fact, the IoT industry is already strongly betting on this promising technology, proof of which is the variety of emerging enterprises with a large portfolio of WET solutions, for example Powercast, TransferFi and Ossia[2].

2.1.2 RF Energy Transfer

RF–WET systems consist of three key components as illustrated in Figure 2.1:

RF transmitters, which constitute the energy sources. The RF transmitters may be intended (usually called PBs when wireless information transfer

---

2 See https://www.powercastco.com, https://www.transferfi.com and https://www.ossia.com.

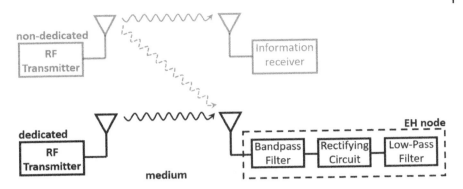

Figure 2.1 RF–WET system.

(WIT) functionality is secondary or simply not integrated), not-intended or a mixture of both, for the specific RF–WET application. The antenna directionality, polarization and transmit power determine the impact of each transmitter on the EH performance, and are in general subject to regulatory and safety standards [95].

Medium encompassing the wireless channels, which are subject to large-scale and small-scale fading, shadowing and blockage impairments.

EH node equipped with an energy conversion circuit consisting of [95]

- A receive antenna(s), which can be designed to work on either single frequency or multiple frequency bands such that the EH node can harvest from single or multiple sources simultaneously. Nevertheless, the RF energy harvester typically operates over a range of frequencies since energy density of RF signals is diverse in frequency.
- A combination matching network/bandpass filter, which consists of a resonator circuit operating at the designed frequency to maximize the power transfer between antenna and rectifier. It ensures that the harmonics generated by the rectifying element are not re-radiated to the environment and that the efficiency of the impedance matching is high at the designed frequency.
- A rectifying circuit, which is based on diodes and capacitors. Generally, higher conversion efficiency can be achieved by diodes with lower built-in voltage. The capacitors ensure that power is delivered smoothly to the load. Additionally, when RF energy is unavailable, the capacitor(s) can also serve as a reserve for a short duration.
- A low-pass filter, which removes the fundamental and harmonic frequencies from the output, sets the output impedance, and stores charge for consumption.

Regarding the EH circuit, the semiconductor-based rectifier is quite common because of its low cost and small-form factor. Specifically, CMOS technology with diode-connected transistors are mostly adopted to significantly increase the EH efficiency at lower powers because of lower parasitic values and customizable rectifiers. This technology is particularly suitable for low-power low-cost applications such as RFID since the entire device can be incorporated on a single integrated circuit (IC). Meanwhile, rectifying antenna (rectenna)-based designs are preferable when larger power densities are available. Readers may refer to [95, 96] for a detailed overview of the main rectenna designs and EH circuit topologies, which include half-wave rectifier, single shunt rectenna, single stage voltage multiplier, Cockcroft–Walton/Greinacher/Villard charge pump, Dickson charge pump, and modified Cockcroft–Walton/Greinacher charge pump.

2.2 RF–EH Performance

After going through the channel, the RF energy is harvested at the EH node. The EH circuitry is characterized by a function $g : \mathbb{C} \mapsto \mathbb{R}^+$ mapping the incident RF signal to the corresponding harvested DC power. We will focus mainly on the non-decreasing EH transfer function with the incident RF waveform power as sole input (i.e., given a certain type of waveform), thus, written as $g : \mathbb{R}^+ \mapsto \mathbb{R}^+$; although later in Section 7.4 we discuss the impact of the waveform type and efficient designs.

With the EH transfer function in place, the power conversion efficiency (PCE) given by $\eta(x) = g(x)/x$ is a KPI, always specified in a circuit's data sheet. The PCE depends on the efficiency of the antenna, the accuracy of the impedance matching between antenna and rectifier and its power efficiency. Additionally, sensitivity and saturation levels are of great interest, and can be obtained from the PCE information. Specifically, the sensitivity is defined as the minimum input power necessary to excite the EH circuit, $\{\min x : g(x) > 0\}$; while the saturation refers to the required level for which the diode starts working in the breakdown region, and from that point onward the output DC power keeps practically constant, mathematically $\{\min x : g(x) \approx g(x + x_0), \forall x_0 > 0, g(x) > 0\}$.

2.2.1 Analytical Models

Usually, the power conversion efficiency, g, is assumed to be linear for analytical tractability, thus, independent of the input power. However, in practice the PCE actually depends on the input power and consequently, the relationship between input and output power is non-linear because of the non-linear RF-DC rectifier behavior. Table 2.2 overviews the main models used for characterizing the EH transfer functions in the research literature. A model's accuracy

Table 2.2 Analytical EH transfer functions.

Model	$g(x)$	Fitting Region
Linear	ηx	in small RF power ranges between sensitivity and saturation levels
Above-sensitivity [76]	$\mathbb{I}(x \geq \varpi_1)\eta x$	similar to Linear but considering the sensitivity phenomenon
Piece-wise linear [84]	$\mathbb{I}(\varpi_1 \leq x \leq \varpi_2)\eta x$ $+\mathbb{I}(x > \varpi_2)\eta\varpi_2$	similar to Linear but considering the sensitivity & saturation phenomena
Quadratic [97]	$a_2 x^2 + a_1 x + a_0$	close to saturation
Fractional [98]	$\dfrac{b_2 x + b_1}{x + b_0} - \dfrac{b_1}{b_0}$	adjustable above sensitivity
Cubic Fractional [99]	$\dfrac{q_3 x^3 + q_2 x^2 + q_1 x}{p_3 x^3 + p_2 x^2 + p_1 x + p_0}$	adjustable above sensitivity
Sigmoidal [100]	$\dfrac{\varpi_2(1 - e^{-c_1 x})}{1 + e^{-c_1(x-c_0)}}$	above sensitivity

η, $\{a_i\}$, $\{b_i\}$, $\{c_i\}$, $\{q_i\}$, $\{p_i\}$ are constants determined by standard curve-fitting and are related to the detailed circuit specifications such as the resistance, capacitance, and diode turn-on voltage. Specifically, η is a constant PCE possibly suitable for the region of interest. ϖ_1 and ϖ_2 are the sensitivity and saturation RF levels. Meanwhile, $\bar{\varpi}_2$ is the EH saturation level, i.e., $\bar{\varpi}_2 = g(\varpi_2)$. Notice that for the fractional and cubic fractional models, one can set $\{b_i\}$, $\{q_i\}$, $\{p_i\}$ such that $b_2 - b_1/b_0 = \bar{\varpi}_2$ and $q_3/p_3 = \bar{\varpi}_2$, respectively, to fit to the saturation region.

depend on the fitting regions and on the specific characteristics of the EH circuits.

The sigmoidal model presented in [100] is a widely adopted non-linear model. However, although it accounts well for relatively moderate to high input RF powers, it fails to emulate the sensitivity phenomenon. A cubic fractional model was later proposed in [99] after the authors examined dozens of practical energy harvesters; while in [98], a simplified, more suitable for statistical analysis, fractional model was alternatively suggested. In both cases, the models are not suitable for mimicking the sensitivity phenomenon, at least without considerably affecting the fitting in other operational regions. Meanwhile, the quadratic model in [97] is an over-simplified non-linear model that fits regions close to saturation and allows some tractable analytical analysis but deviates significantly from real measurements in other regions. The model proposed in [76] comes from an attempt to take advantage of the simplicity of the commonly used linear model while still considering the sensitivity phenomenon. This model was later improved by incorporating the saturation effect in [84]. Therein, authors consider a piece-wise linear function that allows a good trade-off between analytical tractability and fitting to real measurements. Notice that in the region between sensitivity and saturation, non-linear analytical functions can also be adopted, as in [101], for a refined fitting, although over-complicating the subsequent analysis and hindering possible analytical insights.

In a nutshell, there is no one-size-fits-all model, and the trade-off between accuracy and tractability needs to be evaluated case by case before adopting an appropriate approach.

2.2.2 State-of-the-art on RF EH

Table 2.3 compiles the state-of-the-art (from 2015 to 2020) EH circuits, which are diverse in terms of fabrication (antenna and rectifier technology, and dimensions), operation frequency and bandwidth (sub-6GHz and mmWave with narrow and wide-bands), sensitivity and PCE levels.

Note that each of these state-of-the-art designs fit certain applications, for example, ranging from pure ambient RF EH in big cities to space-to-space dedicated WET. For ambient RF EH, multi- and wide-band designs are key to ensure harvesting usable amounts of energy. In these cases, EH circuits with low sensitivity levels and high efficiency in the low-power regime are more convenient. Meanwhile, single-band or tunable EH circuits with high-gain antennas and greater sensitivity levels may be preferable for dedicated WET applications.

As an illustrative example, consider the circuit proposed in [119] for ambient RF EH. Therein, the authors present an adjustable circuit design that can operate on multiple different cellular (LTE 700 MHz, GSM 850 MHz) and ISM bands (900 MHz) with one single circuit. In Figure 2.2(a), we depict the RF input versus harvested power of their circuit design at 700 MHz along with the fitting offered by some of the analytical models presented in the previous subsection. As observed, the proposed circuit is highly sensitive and allows operation with a wide range of input powers without saturating. Saturation occurs around 10 mW, which is very unlikely to be reached in ambient EH applications. Note that the tested analytical models fit quite well in the region 0.1–10 mW (and also between 0.001–0.1 mW, which is not shown here to ensure good readability of the figure), while the sigmoidal and even the piece-wise linear models succeed also in accurately mimicking the saturation region. On the other hand, we also discuss the circuit proposed in [104] for dedicated EH. Therein, the authors presented a compact reconfigurable rectenna for dual band (5.2 and 5.8 GHz) operation. As noticed in Figure 2.2(b), the working input power region is much more compact this time with around 5 mW of margin between sensitivity and saturation. For these operating characteristics, the models' fitting is much less precise. Only the piece-wise linear model completely accounts for the sensitivity phenomenon, while the remaining sigmoidal and quadratic approaches fit the region above sensitivity well, including saturation. The linear model is quite loose in this case. Once again, all this corroborates our claim in the previous subsection that there is not a one-size-fits-all model.

Table 2.3 State-of-the-art (2015-2020) EH circuits in published works.

Ref.	Antenna	Rectifier	Gain (dBi)	Freq. (GHz)	Dimens. (mm)	ϖ_1 (μW)	PCE−η
[102][2015]	microstrip	on-chip CMOS	5.3	160	0.5 × 0.22	100	8.5% @0.57 mW
[103]	patch	volt. doubling	3 – 9	1.9 – 3.2	40 × 40 × 0.8	320	70% @0.8 mW
[104]	patch, metal strip	shunt Schottky reconfigurable	(i) 4.33, (ii) 6.64	(i) 5.2, (ii) 5.8	68 × 34	10^4	30% – 65%
[105]	stack differential	half-wave	6.6	5.8	120 × 40	24	44% @200 μW
[106]	cross dipole	full-wave Greinacher	2.5 – 4	1.8 – 2.5	70 × 70 × 13	0.32	55% @0.1 mW
[107][2016]	i) air-substrate patch, (ii) printed CoCo array	single series Schottky diode	7	2.45	(i) 58 × 10, (ii) 261 × 5	—	30% @1 μW/cm^2
[108]	polarization patch	shunt Schottky diode	4.5 – 6	(i) 5.1 – 5.8, (ii) 5.8 – 6.1	90 × 160	—	(i) 23.8% – 31.9%, (ii) 22.7% – 24.5%
[109]	patch	differentially-driven	5.5	2.45	100 × 70	10	74% @0.2 mW/cm^2
[110]	dual-linearly polarized patch	Schottky diode dual-input	7.5	2.45	70 × 47.5	10	78% @0.3 mW/cm^2
[111]	microstrip patch	Schottky differential	(i) 2.1, (ii) 4.2	(i) 0.9, (ii) 1.8, 2.1, 2.4	115 × 52 × 1.6	10	86% @0.1 mW, 24% @10 μW
[112]	planar monopole	—	5	2.4 – 18.6	25 × 25 × 1.2	—	< 82% @10GHz
[113][2017]	off-center-fed dipole	single shunt diode	(i) 1.8, (ii) 3.4	(i) 0.9 – 1.1, (ii) 1.8 – 2.5	100 × 50	1	≤ 70%
[114]	broadband hybrid dielectric resonator	Schottky HSMS-285C	≤ 9.9	1.8 – 3.6	—	10	< 62, 61, 54, 44% @1.8, 2.1, 2.4, 3.6GHz
[115]	inset-fed patch	Greinacher volt. doubler	8.76	5.8	—	100	< 74.4%
[116]	printed bow-tie dipole	volt. quadrupler	—	1.8	88 × 80	—	—

Table 2.4 State-of-the-art (2015-2020) EH circuits in published works. (continuation)

Ref.	Antenna	Rectifier	Gain (dBi)	Freq. (GHz)	Dimens. (mm)	ϖ_1 (μW)	PCE$-\eta$
[117][2018]	microstrip	HSMS2850 low-barrier Schottky	(i) < 9.3, (ii) 4 − 6	(i) 0.8 − 1, (ii) 1.7 − 2.7	110 × 110	8	(i) 55%, (ii) 35%, @0.7mW
[118]	microstrip multiport pixel	single series Schottky diode	~ 3	1.84	98 × 56	0.1	21% @10 μW, 7% @1 μW
[119]	log-periodic PBC	one-stage Dickson	6	0.7, 0.85, 0.9	—	1	< 45%
[120]	differentially fed multiband slot	Villard volt. doubler	(i) < 7, (ii) < 5.5, (iii) < 9.2	(i) 2.1, (ii) 2.4 − 2.5, (iii) 3.3 − 3.8	160 × 160	3	(i) < 55%, (ii) < 30%, (iii) < 15%
[121]	microstrip patch	4-stage volt. doubler	(i) 5.5, (ii) 6.3	(i) 2.4, (ii) 5.8	41 × 44	300	45% @10 mW
[122][2019]	microstrip PIFA	single/double stage Dickson	< 4	2.4, 3.9	36 × 28	0.01	—
[123]	Hilbert fractal shaped microstrip	Villard volt. doubler	2.2	0.9	80 × 82 × 1.5	0.1	80% @180 μW
[124]	circular microstrip	series-shunt Schottky	3.8 − 9.3	1.85 − 1.93, 2 − 2.1	170 × 170 × 170	1	< 53.6%
[125]	dual band patch	volt. doubler	(i) 2, (ii) 5	(i) 1.8, (ii) 2.5	39 × 40	< 1	(i) < 60%, (ii) < 46%
[126][2020]	patch	2-stage Dickson's charge pump	—	0.915	—	—	70%
[127]	microstrip patch	volt. multiplier	5	5	25 × 27	< 1	< 46%
[128]	3 × 5 MTM Hilbert patch	9-stage Villard volt. doubler	(i) 1.2, (ii) 2.8, (iii) 4	(i) 3, (ii) 5.8, (iii) 7.5	15 × 32	—	90% @0.1 mW
[129]	substrate-integrated waveguide	graphene FET	8.12	29 − 46	3.2×13.2×0.4	< 1	80% @1 mW
[130]	monopole microstrip	volt. doubler	—	0.9, 1.8, 2.5, 5.2	48 × 42 × 1.6	10	< 71.1%

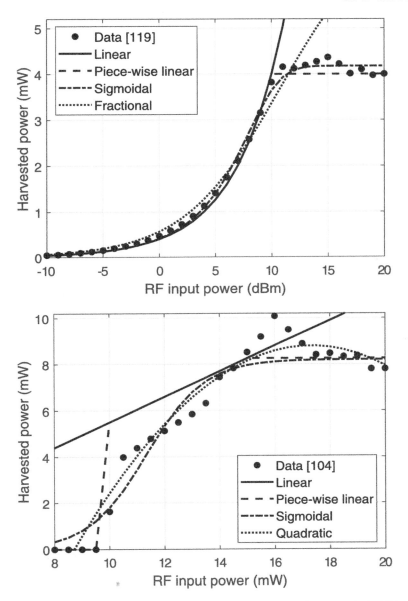

Figure 2.2 Harvested power versus RF input power. Fitting of some analytical EH transfer functions for the EH circuit (a) Data [119] @700 MHz (top), and (b) Data [104] @4.9 GHz (bottom). We set $\eta = 0.4$, $\varpi_1 = 1~\mu$W, $\varpi_2 = 10$ mW, $b_2 = 9.21$, $b_1 = 2.8$, $b_0 = 4.2$, $c_1 = 0.32$, $c_0 = 2.8$ in a); while $\eta = 0.55$, $\varpi_1 = 10$ mW, $\varpi_2 = 15$ mW, $a_2 = -0.12$, $a_1 = 4.1$, $a_0 = -26.7$, $c_1 = 0.95$, $c_0 = 11.3$ in b).

2.3 RF–EH IoT

Herein, we describe the general architecture of an RF–EH IoT network, and the main layouts for WIT and WET along with potential applications.

2.3.1 Architectures of IoT RF EH Networks

The IoT paradigm intrinsically includes WIT, hence RF EH appears naturally combined with WIT. Typical infrastructure-based/-less architectures for an RF–EH network are illustrated in Figure 2.3.

In an infrastructure-based architecture, there are three major components:

- the information gateways, which are generally known as BSs, wireless routers and relays;
- the RF energy sources, which can be either dedicated RF energy transmitters (the so-called PB) or ambient RF sources (e.g., TV towers) as discussed in Section 2.1.1;
- the network nodes/devices, which are the user equipments (UEs) with/without EH capabilities and which may communicate with the information gateways.

Typically, the information gateways and RF energy sources have continuous and fixed electric supply, while the network nodes harvest energy from RF sources to support their operations. In some cases, the information gateway and RF energy source can be co-located or even be the same, i.e., the so-called hybrid PBs/BSs.

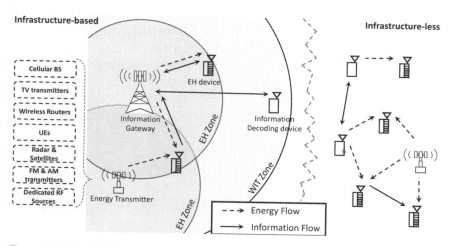

Figure 2.3 General architectures of RF–EH networks. *Source:* Adaptation from [131, Fig. 2].

The EH and WIT zones are illustrated in Figure 2.3. The devices in the EH zone of the information gateway are able to harvest RF energy, while the devices in the WIT zone can successfully decode information transmitted from the gateway. Note that WIT zones are greater than EH zones since the circuitry for energy and information transmissions operate with very different sensitivity levels. While typical information receivers can operate with sensitivities ranging from -130 dBm to -60 dBm receive signal power; an EH device needs usually more than -30 dBm as shown in Table 2.3. The reason is that for information decoding (ID) the metric of interest is based on a ratio, e.g., the SNR or signal-to-interference-plus-noise ratio (SINR), and what matters is how much stronger/weaker the signal power is compared to the noise (+interference) level; while for EH purposes, the nominal received power is what matters.

In the infrastructure-less architecture, on the other hand, there is no infrastructure-based information gateways, and WIT occurs peer-to-peer. Some peer-to-peer WIT transmissions may be important sources of RF energy for nearby EH devices, which may also be served by ambient and dedicated RF energy transmissions to promote sustainability, as discussed next.

2.3.2   Green WET

WET is undoubtedly a sustainability promoter. However, the usual conception of WET-enabled networks, where PBs and hybrid BSs are reliably powered by traditional energy sources, e.g., carbon-based energy, is not strictly compatible with the vision of a fully sustainable IoT. Instead, WET becomes greener by powering PBs and hybrid PBs using energy harvested from renewable sources [132, 133]. This can be implemented in two main ways [78]:

- passively, where no locally-available green energy sources are exploited. The green energy is harvested at the industry side and distributed to the WET transmit infrastructure, which requires specific green contracts with the energy provider; or
- actively, where the green energy is harvested locally by the equipment of the WET provider. For example, PBs connected with outdoor solar panels may use such energy to wirelessly recharge indoor RF-EH devices.

Note that other dedicated WET technologies, e.g., laser power beaming, can be incorporated and serve as a bridge to transport the energy from the green source(s) to PBs.

Among above variants, active green WET is the most attractive as it exploits local renewable sources to strictly realize a small-scale sustainable ecosystem by itself. However, it comes with the typical energy availability challenge of ambient EH. To gain in performance robustness against green energy shortages, the aforementioned variants may be combined forming a hybrid, or intelligent energy balancing [133] and trading [134] mechanisms between PBs and hybrid BSs may be introduced, e.g., to distribute surplus energy properly.

Holistic optimization frameworks must consider the fluctuations in energy availability in the short and large time-scale to fully leverage green WET as a sustainable, efficient and scalable technology.

2.3.3 WIT-WET Layouts

The two main layouts in which WIT and WET appear combined are:

- Wireless-powered communication network (WPCN), where WET occurs in one direction, e.g., downlink in an infrastructure-based architecture, in a first phase, and WIT takes place second in other direction. The energy transmitter may or may not be the same as the information receiver. Some WPCN scenarios of interest and related discussions are presented in Chapter 6.

  Note also that separate wireless communications enabled by either dedicated or ambient RF EH fall into this category, as well as the so-called backscatter communications, which enable the devices to communicate through scattering incoming (ambient or exciting) RF signals;
- Simultaneous wireless information and power transfer (SWIPT), where WET and WIT occur simultaneously in the same direction, e.g., downlink in an infrastructure-based architecture. SWIPT scenarios are discussed in depth in Chapter 7.

2.3.4 RF EH in IoT Use Cases

RF–EH is a promising approach for partially or completely powering many IoT use cases as illustrated in Figure 2.4. Although, WET appears intrinsically combined with WIT as discussed in the previous subsection, we would like to highlight that WET constitutes a fundamental and sensitive building block in IoT networks since its duration could be significantly larger than WIT in order to harvest usable amounts of energy [78]. In fact, some use cases require almost permanent WET operation while WIT happens sporadically, e.g., due to event-driven traffic.

Applications for RF EH in the IoT era are extremely diverse, and affect the circuits' design philosophy. Optimizations for one application may be detrimental to another, sacrificing input power range for PCE, size for durability, or complexity for cost in ways suited only to a specific task [135]. Thus, an application's requirements must be fully considered to ensure circuit trade-offs are given proper weight in the final product. Next, we overview some specific applications for RF EH in the IoT era.

RFID. RFID is a key technology in identification, tracking, and inventory management distributed applications. By incorporating RF EH to the RFID tags, the lifetime and operation range is considerably extended compared to traditional designs [131]. The system operation requires the RFID transceiver to receive data, and also transmit it passively via reflective backscatter managed

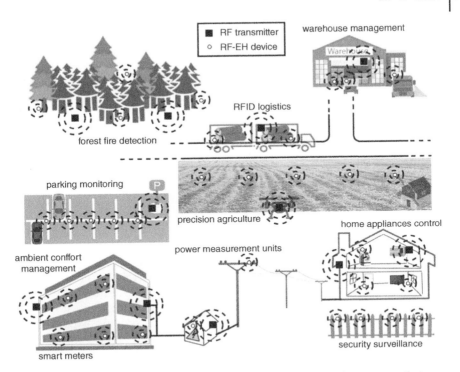

Figure 2.4 RF EH for IoT use cases, e.g., smart thermostats and HVAC (heating, ventilation, and air conditioning) monitoring and control systems that enable smart homes; transportation, healthcare, industrial automation, emergency detection and response, precision agriculture, among others. *Source:* Adaptation from [78, Fig. 6].

by an internal oscillator and signal modulator, or actively by consuming stored energy resources [135]. The latter allows features such as sensing, on-tag data processing and intelligent power management to be introduced, but at the cost of greater complexity and form factor. To save power and reduce hardware complexity/cost, RF-powered RFID chips tend to have fewer rectifier stages for efficiency, small and inefficient antennas, more complex impedance matching networks to provide backscatter, little to no power buffering to conserve space, and minimal computing power [135].

Medical and Healthcare. Applications of RF–EH in medical and healthcare are growing fast. In case of on-/in-body human applications, dedicated RF–EH is not suitable/permitted as the effects of considerable amounts of RF energy impacting on the human body are not yet well understood, instead ambient RF–EH can be adopted for ultra-low power applications. In fact, the first wireless battery-free bio-signal processing system on chip was introduced by [136]. This system was able to monitor various bio-signals via electrocardiogram, electromyogram, and electroencephalogram. The total size of the chip was 8.25 mm^2 and consumed 19 μW to measure heart rate. The power module of this system comprised an RF–EH that was supported by a thermoelectric generator, which together supplied a voltage of 1.35 V. Other attempts for RF–EH-enabled medical applications in the literature include the work in [137], where the authors designed a double-resonant coil ferrite rod medium-band antenna for RF EH in health monitoring systems, and showed its performance remarks. Meanwhile, very recently the authors of [138] designed an EH protocol that harvests energy from two ambient energy sources, namely RF at 2.4 GHz and thermal energy. They showed the considerable lifecycle improvements of the medical IoT devices powered by such sources. Other low-power sensors than can benefit from RF–EH are those utilized for measuring intraocular pressure, bio-impedance, thoracic impedance variance, temperature, and pressure [139].

Wireless Sensor Networks. Wireless sensor networks (WSNs) are at the core of the IoT, covering applications from smart houses and healthcare to industry and military. In recent years, WSNs have gained widespread popularity along with the development of the micro-electromechanical systems (MEMS) technology. These trends will continue and intensify as discussed in Chapter 1. Applying RF energy to recharge or avoid the need of batteries is one promising approach to enhance the lifespan of WSNs, and it is being widely considering as demonstrated by the wide variety of RF–EH sensors available nowadays (see Table 2.3). For instance, [140] conceptualizes the use of batteryless devices in a WPCN together with UAVs as an efficient solution for monitoring remote areas, particularly for maritime applications. Such a solution enables dynamic deployment of infrastructure-less monitoring networks. The authors in [141] proposed their application for structure monitoring of buildings. Specifically, far-field RF–EH sensors were developed to detect humidity, temperature and light inside a building. Results showed that at a distance of 1 m from a 3 W source, the sensor node received 3.14, 2.88, 1.53, 0.7 mW power through air, wood, 2 inches of brick and 2 inches of steel, respectively, which is sufficient for powering many state-of-the-art sensors. Obviously, in many cases, and as the previous example suggests, there is the need for dedicated nearby energy transmitter(s), PBs, to support the applications. In another set of applications, passive RF–EH systems may assist WSN nodes with greater energy demands [135]. When the harvested energy is not sufficient to completely charge the sensors, a passive chip may still be used as a wake-up radio (WUR) that generates a wake up pulse upon receiving a command from a nearby transmitter. Note that the use of a fully passive WUR means that the active portion of the sensor will only be active for short periods, and does not need to actively listen for commands during downtime, which certainly improves the sensor lifetime.

2.4 Enabling Efficient RF-WET

Enabling efficient RF EH is mandatory for realizing the green IoT paradigm but will still encounter many challenges in the future [142], for instance: 1. increasing the end-to-end system efficiency; 2. supporting ubiquitous energy accessibility within the network coverage area, possibly with QoS guarantees; 3. resolving the safety and health issues of WET systems and compliance with regulations; 4. limiting further the energy consumption of EH devices; and 5. seamless network-wide integration of wireless communication and energy transfer.

In addition to these challenges, there will be a fundamental need for powering more and more devices each time, therefore efficient techniques for powering massive IoT deployments are required. Many of these challenges are addressed in the next chapters in different network scenarios.

Note that in the case of ambient RF EH, the action set is limited, and thus relies on the continuous advances of EH circuitry and reduced circuits' power consumption, and on adopting energy-efficient policies and/or on the appropriate deployment of the EH devices such that their functioning is guaranteed as expected. Meanwhile, in case of dedicated WET, the action set not only encompasses the previous strategies but also some other important actions, many of which are on the network side. Next, and based on the discussions in [78], we overview several of these promising techniques.

2.4.1 Energy Beamforming

Energy beamforming (EB) allows direct energy beams to be transmitted toward the end devices, which improves end-to-end efficiency, and is accomplished by carefully weighing the energy signals at different antennas such that a constructive superposition is attained at the intended receivers [143]. The number of antennas not only limits the number of beams, but a larger number allows the focus of the energy beams in some specific spatial directions to be sharpened. However, accurate CSI, which is often difficult or too costly to acquire in WET systems, is required for precise EB. This reveals the need in many cases of CSI-limited schemes as highlighted next.

2.4.2 CSI-limited Schemes

CSI-based strategies such as precise EB are often too costly for RF–EH systems. On one hand, acquiring the CSI via sending training pilots and waiting for a feedback from the EH devices is not desirable since [78]:

- many simple EH devices do not have the required baseband signal processing capability;
- significant amounts of time and energy are required, which may erase (or even reverse) the gains from EB; and moreover,
- the problem of reliable CSI feedback persists.

Alternatively, it seems appropriate that the EH devices send training pilots, while the transmitter estimates the CSI and forms the energy beams. This reverse-link training becomes suitable when using large antenna arrays, e.g., mMIMO or large intelligent surfaces (LISs), since training overhead is independent of the number of transmit antennas. However, accurate channel reciprocity must hold, even though it is sensitive to hardware impairments, especially when devices at both link extremes are very different. Additionally, the EH devices still need carefully designed training strategies, such as the transmit power, duration, and frequency bands, to minimize the energy spent in scheduling and sending the training signals. Notice that an efficient MAC orchestration is necessary to avoid pilot collisions and consequent extra energy expenditure in the collision resolution phase. In general, a basic time-division multiple access (TDMA) or frequency-division multiple access (FDMA) scheme is not appropriate, and much more evolved MAC schemes are required to account for the different (and variable) energy requirements of the EH devices, and the QoS demands of both data transmission and energy transfer in WPCNs. For some scenarios, there is also the problem of receiver mobility which could lead to time-varying channels that makes channel tracking difficult. In both downlink or uplink training, energy/time limits CSI acquisition procedures, which potentially produce substantial errors in estimation and quantization.

Most importantly, the need to support massive IoT deployments poses important challenges not only for the access but for the CSI acquisition procedures (in addition to those previously mentioned). To prevent interference and collisions during training, scheduling strategies may be necessary, draining important energy resources that are costly for energy-limited devices. Moreover, the performance of CSI-based systems decays quickly as the number of users increases [84]. Therefore, intelligently exploiting the broadcast nature of wireless transmissions in such massive deployment scenarios is of paramount importance, even more so when powering a massive number of devices simultaneously with minimum or no CSI.

Instead of EB based on instantaneous CSI, one may exploit statistical knowledge of the channel, so-called partial CSI, which varies over a much larger time scale and consequently is learned with limited CSI-acquisition overhead and energy expenditure. Partial CSI-based EB schemes are particularly beneficial in setups where the PB is a typical mMIMO or LIS node and downlink/uplink channels are not reciprocal. The benefits are surely promising in future network deployments with IRSs (see Chapter 1), since an IRS may assist blocked energy transmissions by providing alternative passively controlled LOS paths. CSI estimation/acquisition in this kind of deployment remains a challenge, while partial CSI-based EB could attain near optimum performance since some LOS is expected between transmitter and IRS, and between IRS and EH devices, which strongly reduces channel uncertainties.

2.4.3 Distributed Antenna System

End-to-end efficiency of EH systems decays quickly with distance, following a power-law, between the energy transmitter and the EH device. In this context, DASs become important enablers of future IoT networks because of their strong potential to eliminate blind spots, while homogenizing the energy provided to a given area and supporting ubiquitous energy accessibility. Allowing each separate multi-antenna transmitter to take responsibility for powering a smaller set of EH devices can potentially alleviate the CSI acquisition issue [144].

Future WET systems will undoubtedly benefit from novel kinds of DAS deployments such as cost-efficient radio stripes systems [145]. In these systems, the actual PBs will consist of antenna elements and circuit-mounted chips inside the protective casing of a cable or stripe. While traditional antenna deployments may be bulky, radio stripes enable invisible installation in existing construction elements and alleviate the problem of deployment permissions. Besides, cables are malleable, and the overall system is resilient to failures because of its distributed functionality. However, optimized resource allocation schemes, circuit implementations, prototypes, and efficient distributed processing architectures to avoid costly signaling between the antenna elements, are still necessary. Figure 2.5 illustrates EB, pure DAS, and a more attractive combination of these two, either via traditional or more novel kinds of deployments.

2.4.4 Enhancements in Hardware and Medium

Ultra-low-power Receivers. Receive architectures with ultra-low power consumption are essential to efficiently enable WET to a variety of future IoT applications. The incorporation of power management units (PMUs) may be key to realize ultra-low-power receivers. These circuit blocks monitor the harvested energy levels and provide the charge control/protection of the energy storage units such as capacitors or batteries [146]. However, the intelligence that can be deployed into a PMU is limited by its allowed energy consumption. In general, significant power reduction/efficiency gains can still be achieved in the coming years by optimizing the EH hardware as a whole, e.g., by integrating antennas/rectennas into the device package with an optimized intimate connection [35], and designing/integrating PMUs that optimize the net harvested energy (harvested energy minus power consumption over time) to ensure real system gains. Additionally, WURs, duty cycling or event-driven architectures (with short power-up settling times to enable swift change between sleep and on states) may be adopted to reduce transceiver usage.

Enhanced PB. The traditional fixed PB concept is evolving to more efficient and flexible implementations, e.g., rotor-equipped PBs [85, 87],

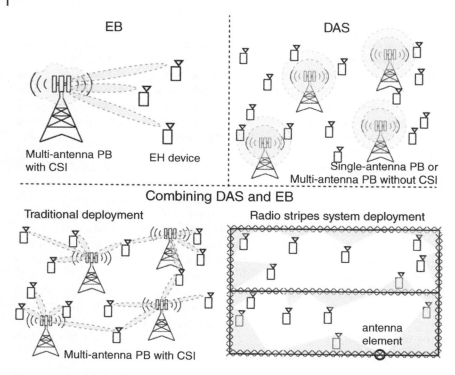

Figure 2.5 Enabling efficient WET via EB, DAS and EB & DAS. *Source:* Adaptation from [78, Fig. 8].

motor-equipped PBs [147–149] and PBs mounted on UAVs [150]. Specifical-ly, the WET efficiency improves by properly rotating the PB, specially under the influence of some LOS. Meanwhile, motor-equipped PBs are particular-ly attractive for periodically or on-command wandering around and power their service area. Mobile charging requires fewer PBs than fixed PB deploy-ments, and is more adaptable to a dynamic IoT network topology as charging routes can be easily re-adjusted by a centralized network planner entity or even autonomously by the PB itself via AI mechanisms. Also, by incorporating high-gain directional antennas into the PBs, further considerable improve-ments can be attained. Finally, UAVs may provide additional advantages over ground PBs such as [78] 1. aerial mobility; 2. flexible accessibility to remote, rural or disaster areas where gaining ground access is difficult/impossible; and 3. great chances of reaching favorable positions with respect to IoT devices for downlink WET and downlink/uplink WIT. Therefore, unmanned aerial PBs (UAPBs) can potentially foster RF–EH IoT applications such as environment/ farm monitoring and emergency services, where UAPBs wake up IoT deployments via WET, and collect data. Further advances on reconfigurable antennas [151, 152] and UAV swarms operation constitute potential technological enablers [153] as they allow realizing proper beam footprint adjustments and high transmit gains, which may significantly boost the QoS-based energy coverage of the UAPBs.

Reprogrammable Medium. WET propagation medium can be conveniently "influenced" via the strategic deployment of smart reflect-arrays and reconfigurable meta-surfaces [54]. For instance, IRSs allow on-fly and opportunistic reconfiguration of the propagation environment without additional energy consumption/expenditure, and are considered a key technological enabler of future green networks [53, 54, 78]. Research community has mainly focused on IRS-assisted communication scenarios, but IRS-assisted RF-powered systems has been gaining interest recently, e.g., refer to [154–156]. IRS are beneficial to either compensate the significant path losses thanks to their large aperture arrays, or to provide alternative WET links to obstructed direct paths. It has been shown that the IRS deployments allow important power savings.

Towards the massive adoption of IRS technology, there are still some challenges that need to be carefully addressed, for instance [53, 78]:

- manufacturing high-precision reflective elements requires expensive hardware, which may not be a scalable solution as the number of reflective elements becomes very large;
- the passive reflect EB of IRS, and active EB of PBs and hybrid BS need to be jointly designed for optimum performance, which leads to complicated optimization problems that are hard to be efficiently solved;
- similar to traditional EB (discussed in Section 2.4.1), instantaneous CSI is required to leverage the passive EB performance gains promised so far in the literature. However, in IRS-assisted WET to a large number of low-power IoT devices, the required overhead may be prohibitive, and efficient CSI-limited/free schemes need to be leveraged.

Finally, note that for an efficient WET to a large set of EH devices, it may be necessary to partition the reflecting elements set such that each partition is responsible for reflecting the arriving signal to a certain device or group of devices.

2.4.5 New Spectrum Opportunities

Recently, the mmWave electromagnetic spectrum is being considered by the research community, and even starting to be exploited by industry, for wireless communication applications. This is motivated by [157]: 1. the large bandwidths that remain unexploited in such spectrum regions; 2. mmWave signal propagation is more directive and of shorter-range which is favorable for spectrum re-use in small cells; and 3. shorter wavelengths allow antenna sizes to be reduced, which translates to either smaller form-factors e.g., at the IoT device side, or to being able to pack more antennas, e.g., at the BS side, and attain high directional gains.

The above advantages hold from the WET point of view and align with the vision of massive miniaturized IoT devices' deployments, e.g., smart dust like motes and zero-energy sensors [158]. Note also that the integration/coexistence of WET and WIT may be more easily promoted since interference issues are considerably relaxed. Furthermore, the larger spectrum bandwidth and high antenna gains may allow the PB to effectively transfer larger amounts of energy under LOS conditions. In fact, assuming that the antenna size is larger than the wavelength, the aperture would be proportional to actual physical cross-sectional area, and the received power can be approximately written as [159]

$$P_r = \frac{P_t}{d^2} \frac{A_t A_r}{\lambda^2}, \tag{2.1}$$

where λ is the signal wavelength, d is the channel distance, P_t and P_r represent transmit and receive power, and A_t and A_r represent area of the transmit and receive antennas, respectively. This illustrates that using higher frequencies with the size limitation in transmitter and/or receiver is indeed beneficial.

Now note that since LOS channels are characteristic of many WET applications and are mostly influenced by the antenna array topology and network geometry [160], CSI-based EB may be avoided (at least partially) by exploiting accurate devices' positioning information. However, the problem of imperfect beam alignment must be carefully taken into account. How to overcome NLOS, or even severe signal attenuation, is an even more critical challenge in mmWave WET with limited CSI, and may require exploiting multi/hybrid RATs. Finally, the highly directive energy beams that are typical in mmWave WET may cause the signal intensity perceived in a particular area to be strong enough to harm human health[3]. DAS (discussed in Section 2.4.3) is a promising approach to solve safety issues as smaller path losses need to be compensated by the beam gains, and could even be combined with sensing-based human real-time detection technologies to cease beam-specific WET when it is deemed to be harmful.

2.4.6 Resource Scheduling and Optimization

Extensive research shows that, under intelligent policies, WPCN may achieve a performance, e.g., in terms of throughput [162], comparable to that of a conventional non-WET network. When information and energy transmissions take place in the same network system, scheduling should 1. avoid co-channel interference or limit its impact (e.g., by taking advantage of synchronization

---

3 The radiation power of any device at certain frequency must satisfy an equivalent isotropically radiated power requirement [161].

and SIC techniques to cancel deterministic WET signals in the network), and 2. optimize the overall system performance according to the metric of interest. In practice, real-time information/energy scheduling is a challenging problem because of time-varying wireless channels and the causal relationship between current WET process and future WIT.

Communication and energy scheduling can be performed in the spatial domain when energy and information transmitters are equipped with multiple antennas [163]. For instance, EB can be used by a PB to steer stronger energy beams to efficiently reach certain users while prioritizing their energy demands towards the information transmission phase. Besides, EB and space-division multiple access can be combined with dynamic time-frequency resource allocation to further enhance the system performance in WPCNs. Some other strategies that have been considered in the literature so far are input signal distribution optimization [142], cooperation [143, 164, 165], hybrid automatic repeat-request (HARQ) [166, 167], power control [76, 142] and rate allocation [77].

2.4.7 Distributed Ledger Technology

WET, as a pivotal component of RF–EH networks, is fundamental for realizing the Internet of Energy (IoE) paradigm [168]. Both infrastructure and IoT nodes may trade their energy goods or surplus with other nearby nodes for which they can act as opportunistic PBs. The distributed nature of this trading creates important challenges in terms of security and privacy. For instance, energy theft, which emerges when a non-authorized device harvests considerable amounts of energy for which it has not paid, or potential attacks from malicious IoT devices, such as energy repudiation of reception, and energy state forgery. Distributed ledger technology (DLT)-based solutions, e.g., Blockchain and Holochain DLT, are promising technologies but there are still some challenges such as the large communication overhead, handling massive two-way connections and energy-efficient DLT protocol design, that need to be efficiently addressed. In addition, powering the devices that actually paid the service must be extremely precise in space, time (and other domains), for which novel and efficient EB designs are necessary, for instance, to benefit the legitimate EH devices while providing negligible energy to the non-authorized ones.

2.5 Final Remarks

In this chapter, we overviewed the main EH technologies and their relevant role in present and future IoT deployments. We emphasized on RF–EH, and highlighted its advantages compared to competing EH technologies, say, reduced form-factor implementation and native multi-user support. RF–EH technologies were classified according to the nature of the RF energy source in 1. ambient RF EH, which does not require additional resource/power consumption from the surrounded network infrastructure, but it cannot guarantee demands with QoS, and 2. dedicated RF EH, which is favorable to support applications with more stringent QoS requirements. Additionally, the EH circuits' performance in terms of PCE was analyzed from an analytical perspective by discussing the main mathematical models available in the literature. At the same time, a wide compilation of state-of-the-art EH circuits was presented while briefly discussing their main distinctions.

The chapter also addressed green WET as a means to fully promote sustainability, as well as key use cases, and the main architectures for RF–EH networks in the IoT, say, infrastructure-based/-less, along with their main components. Since the IoT paradigm intrinsically includes WIT, RF–EH appears naturally combined with the former. Two main layouts in which WET and WIT appear combined, namely WPCN and SWIPT, were discussed. Finally, we analyzed several promising techniques for enabling efficient RF–WET in the IoT era. Specifically, we discussed EB and CSI-limited schemes, where the latter emerge as a natural alternative to CSI-based approaches that may be unaffordable for many ultra-low power RF–EH applications. DAS, efficient resource scheduling and DLT schemes were also analyzed as key technologies to enable efficient and green IoT deployments in the coming years.

3

Ambient RF EH

3.1 Motivation and Overview

Ambient RF–EH (also known as RF energy scavenging) techniques exploit available environmental energy such as that freely coming from broadcast TV, radio, WiFi, Bluetooth, and cellular signals. The following classification makes distinction between the different ambient RF sources that may be available [131]:

- Static ambient RF sources, which support energy transmissions that remain relatively stable over time, e.g., TV and radio transmitters. These static sources allow long-term[1] predictability of the energy supply, which favors network deployment planning.
- Dynamic ambient RF sources, which supply time-varying energy over a certain region, either because of fluctuating transmit powers, e.g., a WiFi AP, or mobility, e.g., cellular users. Consequently, dynamic ambient RF–EH should be adaptive and allow operation on a wide frequency range.

The transmit power of the RF sources may vary significantly, from around 1 MW for a TV transmitter, to about 10 W for cellular BSs and RFID systems, to roughly 100 mW for mobile communication devices and WiFi systems [131]. Normally, the power density in a certain area at different frequency bands is small and high-gain multi-band antennas are desirable. Moreover, the rectifier must also be designed for a wide band spectrum. Anyway, ambient RF–EH is appealing for supporting ultra-low power IoT applications.

The feasibility of ambient RF–EH has been thoroughly investigated in the literature. For instance, in the case of EH from static ambient RF sources, the authors in [169] show that a TV tower can supply 60 μW to devices 4.1 km away. Also, at 6.3 km from the Tokyo Tower, the PCE was measured to be

---

1 In a shorter term there are fluctuations, e.g., due to service schedule and channel fading.

Wireless RF Energy Transfer in the Massive IoT Era: Towards Sustainable Zero-energy Networks. First Edition. Onel Alcaraz López and Hirley Alves.

about 16%, 30%, and 41% when the input power is -15 dBm, -10 dBm, and -5 dBm, respectively [170]. Meanwhile, the study in [171] is an example of dynamic ambient RF–EH in a cognitive radio network. Therein, a secondary user can harvest RF energy from the transmissions of nearby primary users, and can transmit data when it is sufficiently far from primary users or when the nearby primary users are idle. Additionally, the authors of [172] discuss the feasibility of using WiFi RF signals to power several household and personal devices in a 24.5 m^2 area. Interestingly, they show that one WiFi AP is enough to power a small calculator and wall clock, but the APs deployment needs to be extremely dense to power more energy-demanding devices.

2 APs	$\rightarrow$	smoke detector, mouse
6 APs	$\rightarrow$	digital thermometer, wristwatch
38 APs	$\rightarrow$	gas meter,
1774 APs	$\rightarrow$	hearing aid,
84487 APs	$\rightarrow$	lightbulb,
23500 APs	$\rightarrow$	smartphone.

Therefore, ambient RF–EH must be understood as an enabling technology for powering the *ultra-low power* IoT. Table 3.1 summarizes key results from several experiments on EH from ambient RF sources in recent years (note that usually AM and FM transmitters use isotropic antennas). Observe that the EH performance is source, frequency and distance-dependent and the amount of harvested energy is usually in the order of μW.

Table 3.1 Experimental data of RF–EH.

Source	Power	Frequency	Distance	Harvested energy
Isotropic RF transmitter [173]	1.78 W	868 MHz	25 m	2.3 μW
KING-TV tower [174]	960 kW	674 – 680 MHz	4.1 km	60 μW
WiFi AP [172]	100 mW	2.4 GHz	3.3 m	0.5 μW/cm^2
Isotropic RF transmitter [175]	500 mW	900 MHz	1.7 m	42 μW
AM transmitter [176]	50 kW	1.27 MHz	2.5 km	0.6 μW
AM transmitter [177]	10 kW	1 MHz	2.5 km	62 μW
Cell phone [178]	10 mW	ISM bands	1 m	10 μW

3.1.1 Hybrid of RF–EH and Power Grid

As discussed in Chapter 2 and noted previously when observing the ultra-low levels of energy available for harvesting without dedicated RF sources, ambient RF-EH alone is in general not suitable for powering devices with somewhat stringent energy requirements. Additionally, dedicated WET may not be available or may be still insufficient in certain scenarios, which has paved the way for an alternative solution relying on hybrid schemes of RF–EH and power grid [179]. In such a hybrid setup, the energy comes from an energy harvester and a power grid or traditional battery. Energy efficient scheduling algorithms that minimize the energy consumption of the power grid, while ensuring the service requirements, are necessary. For instance, a delay optimal scheduling problem is proposed in [180] for a communication node powered by an EH battery of finite capacity together with a power grid subject to an average power constraint. Therein, the authors show that the communication node will resort to the power grid only when its data queue length exceeds a threshold and no harvested energy is available.

Toward 6G, we may expect very dense deployments of small-cell BSs (SBS)[2] with smaller coverage but requiring less transmit power, thus, they may partially rely on EH techniques. Indeed, the energy consumption from power grids can be effectively reduced by equipping BSs with (different) EH modules [181]. Since relying completely on ambient RF–EH (and other EH sources) is not possible because of the low harvested energy and energy arrival rate uncertainty, each SBS still requires grid connections to avoid severe energy/transmission outages. A recent overview [179] discusses that efficient mechanisms for resource allocation, user scheduling, cell planning, and addressing the doubly near-far problem (RF–EH cellular users at the cell edge can harvest less energy in downlink but require higher transmit power in uplink, which makes the fairness among users challenging) are necessary. Also, adaptive mechanisms for on-grid, off-grid, and idle modes operation are crucial. As highlighted in [182], there could be 1. a central entity optimizing the proportion of on-grid, off-grid, and idle SBSs in the network, and notifying the decision to the SBS, 2. a fully distributed optimization allowing each SBS to decide its operation mode individually in order to enhance its own utility, thus, reducing the signaling/information overhead of 1. which is not scalable for ultra-dense deployments, at the cost of some degradation of the overall system utility; or 3. a semi-distributed mode selection combining the benefits of previous alternatives by clustering the SBS to operate with 1. or 2.

---

2 Wireless networks are getting denser, and this trend is expected to continue in order to sustain the increasingly stringent QoS requirements.

3.1.2 Energy Usage Protocols

Three key protocols for the harvested energy usage have been identified in the literature [179]:

- Harvest-use (HU): There is no buffer to store the harvested energy for future use. The devices perform the required tasks, e.g., sensing, data transmissions, when a sufficient amount of energy is acquirable to cover the associated processing cost.
- Harvest-store-use (HSU): There is an energy storage unit, and the harvested energy can be used only after it is stored in the buffer at the next time instance.
- Harvest-use-store (HUS): The harvested energy that is temporarily stored in a supercapacitor can be immediately used for some low-energy tasks, and the remaining energy is then transferred to another energy storage unit for later use.

At the ith time instant, let B_i be the amount of energy stored in the energy storage unit; B_{max}, the battery capacity; C_i, the processing cost; E_i, the harvested energy; and D_i, the depleted energy. Also, assume that only a portion $0 \leq \beta_1 \leq 1$ of the harvested energy is charged in the battery, and that $\beta_2 \geq 0$ energy in the battery gets leaked in each time slot due to storage inefficiency. Then, we can summarize the battery state evolution under the above protocols as in Table 3.2. Note that for an Ni–MH rechargeable battery, $\beta_1 \approx 0.7$, and for a supercapacitor, $\beta_1 \geq 0.95$ [179]. Meanwhile, the leakage factor β_2 for a battery is very small, while that for a supercapacitor is larger. To illustrate the expected system performance under each of the above energy usage protocols, consider a simple scenario with exponentially distributed harvested energy per time slot as follows. Note that this may correspond to a setup with a unique non-dedicated RF source with channels subject to Rayleigh fading, and without accounting for non-linearities in the EH processes. Figure 3.1 shows the energy outage probability, measured as the chances that the available energy in the battery is not enough to guarantee a use of D_i energy units. Note that under an energy outage, the device's due tasks are not performed in the

Table 3.2 Key energy usage protocols.

Protocol	Battery state evolution (B_{i+1})	Available energy
HU	−no energy storage unit−	$[E_i - C_i]^+$
HSU	$\left[\left[(B_i - C_i - D_i) - \beta_2 \right]^\times + \beta_1 E_i \right]^\times$	$[B_i - C_i]^+$
HUS	$\left[\left[B_i - [D_i + C_i - E_i]^\times + \beta_1 \left[E_i - D_i - C_i \right]^\times \right]^\times - \beta_2 \right]^\times$	$[B_i + E_i - C_i]^+$

Note that $[x]^\times = \min\left([x]^+, B_{max} \right)$ and $[x]^+ = \max(x, 0)$. *Source:* [179].

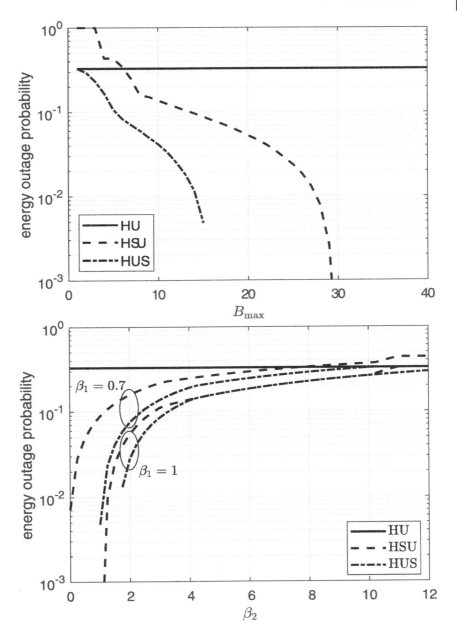

Figure 3.1 Energy outage probability under each energy usage protocol as a function of (a) the battery capacity B_{max} for $\beta_1 = 0.7$ and $\beta_2 = 1$ (top), and (b) β_2 for $B_{max} = 15$ (bottom). We set $C_i = D_i = 2$, $B_0 = 0$ and $E_i \sim \text{Exp}(10)$.

corresponding time slot. As evidenced, HU is often at a clear disadvantage as it just relies on the instantaneous harvested energy, while HUS always attains the best performance since it is less sensitive to battery impairments than HSU. In fact, both HUS and HSU perform the same under ideal battery conditions, i.e., $\beta_1 = 1$, $\beta_2 = 0$ and $B_{max} \to \infty$. However, such conditions, which are often assumed in the literature, do not hold in practice, and a distinction on the utilized energy usage protocol is necessary.

3.1.3 On Efficient Ambient RF–RH Designs

Next, we discuss some approaches for enabling efficient ambient RF–EH.

Tunable-band Harvesters. Multi-band operation is especially attractive for ambient RF–EH applications, where the energy available in a certain frequency band may be extremely low. However, multi-band operation often requires multiple circuits to be operating individually in distinct bands. Such designs increase the complexity and cost of the system, thus, may be inappropriate in massive low-power low-cost IoT use cases. Alternatively, a single-circuit prototype with dynamic frequency tuning, for instance harvesting from a single frequency band that can be switched if sensed to perform badly[3], is preferable. However, efficient tunable-band harvesters require adaptive matching networks as well. Moreover, depending on the application, location, instantaneous consumption and operational circuit voltage, the sensor motes that are interfaced with an energy conversion circuit present a variable load impedance value [119]. Most existing circuits are typically optimized for a specific load impedance, thus, not flexible enough to allow a seamless switch to a different sensor mote, application or location. This would require (some times invasive) adjustments of the impedance matching network of the device, which increases the complexity of repair and cost of fabrication, and impacts the generality of interfacing with a large number of different sensors that may typically compose a massive IoT deployment.

Therefore, tunable-band harvesters with dynamically adjustable impedance matching network are highly desirable for ambient RF–EH applications [119] because of their 1. flexibility, since they can easily adapt to different scenarios without circuitry configuration changes; and 2. lower cost than multi-band multi-circuit harvester prototypes. The single-band, multi-band and tunable-band operational modes are illustrated in Figure 3.2, which complements the basic block scheme presented in Figure 2.1.

---

3 For instance, note that for certain locations, cellular BTB may have greater influence than signals at TV broadcast band and vice-versa. Additionally, even for such locations where cellular BTB are stronger during daytime, that may not hold during nighttime.

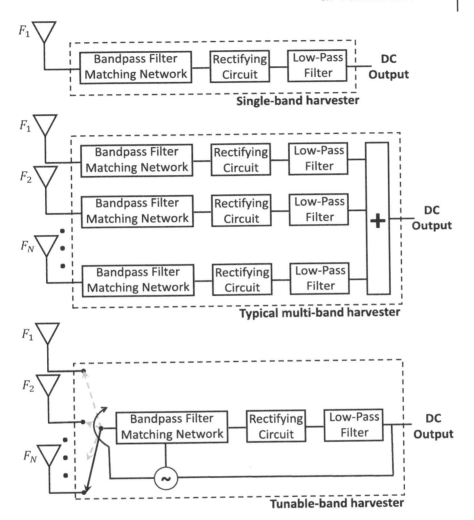

Figure 3.2 Single-band, multi-band and tunable-band RF harvesters.

High-gain Multi-directional Rectenna. Conventional rectenna designs with high-gain compact antennas or antenna and rectenna arrays have been commonly implemented to increase the received RF power. However, the directive nature of high-gain antennas does not fulfill the requirement for the rectennas to be omni-directional for ambient RF EH. Thus, dipole rectennas have been commonly adopted because of their quasi-full spatial coverage of linearly polarized radiations [183, 184]. Strategies like 3D structures with multiple inkjet-printed antennas covering different directions have also been proposed to increase the spatial coverage, e.g., [185]. These antennas exhibit a limited gain nonetheless, yielding a poor rectification efficiency. [186]. In an effort to strengthen the rectifier input power to a higher level, several sources of energy can be collected together and rectified by a single element. In fact, the combination of RF power conducted by antennas with high effective aperture is the most efficient way to increase the receive power. That is why RF power combination and wide beam width radiation are promising and being explored, e.g., in [118, 186–188].

In [118, 188], a power combining rectenna array with a beamforming matrix was proposed to increase both the received power and the beam width in the E-plane of the rectenna. Meanwhile, the authors of [186, 187] proposed and studied the association of five antenna arrays with simple passive beamforming networks directly connected to rectifiers, as shown on Figure 3.3. A passive beamforming network connected to a receiving antenna array controls the phases of incoming signals (at ports 5, 6, 7 and 8) to redirect them into one input port (1, 2, 3 or 4) depending on the beam's orientation (B1, B2, B3 or B4). Therefore, associated to rectifiers (R1, R2, R3 and R4), the system allows RF–EH with high gain antenna arrays without beamwidth selectivity. With enhancement of the power at the rectifiers input, the rectification efficiency is increased in all directions. However, although operational gains are undeniable, much more

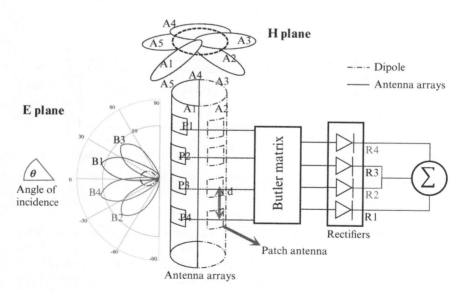

Figure 3.3 Linearly polarized RF power combining a quasi-isotropic energy harvester proposed in [186, 187] and its radiation pattern in the E- and H-planes. The radiation pattern is comparable to the radiation pattern of an omnidirectional electric dipole antenna vertically polarized but with 6 dB higher gain on average. Source: Based on [186].

research effort is still required to further reduce the system size and enable its applicability to future IoT use cases.

Large Harvesting Surfaces. The research community and industry are considering metasurfaces, metamaterial-based antennas and reflectarrays as key in the next generation of wireless systems [189–191]. The associated extremely-low hardware footprint enables their cost-effective embedding in various components of the wireless propagation environment, e.g., building facades and room walls/ceilings, thus, enabling manmade electromagnetic wave propagation control and environmental AI [191]. Although, the focus has been on efficient information transmission/aided wireless networks, we envision early deployments of large RF–EH surfaces in the 6G massive IoT era as well.

Note that by deploying large-area EH antennas or multi-antenna large-surface harvesters, the harvested energy may considerably increase, thus, allowing more energy demanding devices to be powered. Some early advances suggest the viability of such large harvesting surfaces. For instance, a prototype of an optically transparent (microstrip patch) EH antenna on glass was proposed in [192], which can be quasi-invisibly installed in a window without compromising its function. A frequency-selective scalable surface that is specially designed and optimized for ambient RF EH was more recently proposed in [193]. Meanwhile, authors of [194] demonstrate flexible and lightweight textile-integrated rectenna arrays for powering wearable electronic devices by exploiting large clothing-areas. Additionally, RF–EH may be combined with other EH sources to further boost the EH capability [195, 196]. Particularly attractive is the proposal in [195], where authors propose a hybrid rectenna for solar and RF–EH. They demonstrated that the solar cell can be integrated to an RF energy harvester without significantly affecting the performance of the rectenna as long as some design constraints are met. Large-area hybrid solar-RF EH would be very attractive for outdoor deployments, e.g., building facades, where EH was traditionally unavailable during nighttime. Further discussions on technologies enabling the integration of electronics on large surfaces, for instance printing and roll-to-roll compatible techniques, were provided in [196]. Finally, an overview of metamaterials, specifically on their fabrication, exotic properties and achievements in RF–EH, is presented in [197].

3.2 Measurement Campaigns

The feasibility of ambient RF–EH needs to be assessed via on-site RF measurements and/or using simulation tools. RF measurement campaigns offer more fine-grained information about the expected performance but also at greater economic cost. However, while massive measurement campaigns may not be viable, they can be carried out at smaller scale, and used to validate simulation models that can be later extrapolated to other scenarios.

RF measurements, together with proper knowledge on harvester performance, can be used to determine the locations at which RF–EH IoT devices can be successfully deployed. Several RF spectral surveys, which measure ambient RF power levels, have been previously reported, for instance [198–201], some of which originally aim at measuring the personal exposure to RF electromagnetic fields (EMFs), which is crucial for human health and epidemiologic research. Next, we summarize some of the main findings in recent years.

3.2.1 Greater London (2012)

In [198], the authors present the results of a citywide RF spectral survey within Greater London in the frequency spectrum of 0.3–2.5 GHz. Each station on the London Underground network was used as a measurement point, which allowed representative results for the entire Greater London area to be obtained. This is due to the diverse geographical distribution and population density (combining both urban, in the center, and semi-urban, in surrounding areas, environments) of the measurement areas. Specifically, measurements were taken at each of 270 stations during daytime and over a period of one month (March 2012). It is important to highlight that the spectral measurements were undertaken during the analog-to-digital switchover period in the UK, thus, the measurements for Digital TV (DTV) represent a clear lower bound of the RF power levels after such a switch. It should also be noted that the survey was conducted prior to the 4G network being switched on within the UK.

Table 3.3 shows the RF power levels across all London Underground stations for the banded input RF power density measurements[4] S. The GSM-900, GSM-1800 and 3G base transmit bands (BTBs) were separated from the associated mobile transmit bands (MTBs). It can be observed the BTB power levels are between one and three orders of magnitude greater than the associated MTD levels. This is because of the much greater transmit power levels at the network side even when the network node deployments, such as BSs, APs, are less dense.

Finally, authors indicated suitable locations (approximately half of the London Underground stations) and associated RF bands with sufficient input RF power density levels for harvesting using four proposed circuit prototypes. They highlighted the need of multi-band array architectures to provide a broader freedom of operation.

3.2.2 Diyarbakir (2014)

In [199], the authors measured the RF power levels in dense urban, urban and suburban areas of Diyarbakir, Turkey during a whole week in November 2014.

---

4 The banded input RF power density is calculated by summing all the spectral peaks across the band.

Table 3.3 Results of Greater London (2012) RF measurements.

Frequency Band (MHz)		Average S (nW/cm²)	Maximum S (nW/cm²)
DTV	470 – 610	0.89	460
GSM-900 (MTB)	880 – 915	0.45	39
GSM-900 (BTB)	925 – 960	36	1930
GSM-1800 (MTB)	1710 – 1785	0.5	20
GSM-1800 (BTB)	1805 – 1880	84	6390
3G (MTB)	1920 – 1980	0.46	66
3G (BTB)	2110 – 2170	12	240
WiFi	2400 – 2500	0.18	6

Source: [198].

Six different RF bands: FM, TV3, TV4-5, GSM-900 (BTB), GSM-1800 (BTB) and 3G (UMTS-BTB) were measured in the busy hour of mobile network operators in Diyarbakir. Measurements were taken at 1.70 m height above ground on the main streets while moving with average velocity of 40 km/h. From the measurement results, which are illustrated in Table 3.4, it can be observed that cellular BSs provide the greatest power levels. The authors realized that this was because FM and TV broadcast transmitters were far from the streets being measured, whereas some cellular BSs were located nearby. Finally, they highlighted that the measurement values for each band agreed with the admissible EMF levels.

3.2.3 Flanders (2017-2019)

In [201], the authors analyzed the results from RF measurement campaigns carried out between 2017 and 2019 in five of the largest cities in the Flemish Region and the Brussels Capital Region: Antwerp, Bruges, Brussels, Ghent, and

Table 3.4 Results of the Diyarbakir (2014) RF measurements.

Frequency Band (MHz)		Average S (nW/cm²)	Minimum S (nW/cm²)	Maximum S (nW/cm²)
FM	87.5 – 108	2.3	1.3	30
TV3	174 – 230	1.7	1.2	2.6
TV4-5	470 – 862	5.2	3.0	98
GSM-900 (BTB)	935 – 960	61	0.3	3948
GSM-1800 (BTB)	1805 – 1880	13	0.7	2300
3G (UMTS-BTB)	2110 – 2170	100	1.4	9838

Source: [199].

Hasselt. This time, a much wider portion of the RF spectrum was sensed, i.e., 87 − 6000 MHz, comprising 16 frequency bands that were grouped into four categories:

- Downlink: LTE-800 (BTB) 791 − 821 MHz, GSM-900 (BTB) 925 − 960 MHz, GSM-1800 (BTB) 1805 − 1880 MHz, UMTS-2100 (BTB) 2110 − 2170 MHz, and LTE-2600 (BTB) 2620 − 2690 MHz.
- Uplink: LTE-800 (MTB) 832 − 862 MHz, GSM-900 (MTB) 880 − 915 MHz, GSM-1800 (MTB) 1710−1785 MHz, UMTS-2100 (MTB) 1920−1980 MHz, and LTE-2600 (MTB) 2500 − 2570 MHz.
- Broadcast: FM 87.5 − 108 MHz, and DVB-T 470 − 790 MHz.
- Other: DECT 1880 − 1900 MHz, WiFi 2G 2400 − 2485 MHz, WiMax 3.5 3400 − 3600 MHz, and WiFi 5G 5150 − 5875 MHz.

The average power density per category is shown in Figure 3.4. The highest average total power density was measured in Brussels, mainly caused by a relatively high broadcast component. The lowest average total power density was measured in Bruges. Although with a lower population density, the average RF measurements levels were the highest in Antwerp, even compared to Brussels, which is much more densely populated. The authors associated this with the fact that regulatory legislation on BSs are stricter in Brussels. Finally, the authors discussed the representativeness and high repeatability of their measurements, and evidenced the potential impact of different regulations on the population's exposure to RF-EMF radiation.

3.2.4   Other Measurements

In 2008, the authors of [202] had already reported that 109 μW RF power can be harvested from the daily routine in Tokyo. In 2010, the authors of [203] showed that the average power density from 1 GHz to 3.5 GHz is of the order of 6.3 nW/cm^2 in French urban environments. Two different environments in Dhahran (SA) and in Boston (USA) sensed in 2012 showed enough RF activity to excite a rectenna and harvest power through dynamic RF tuning in specific frequency bands within 80–2000 MHz spectrum [204]. In 2016, authors of [204] measured power peaks around −38 dBm at 944 MHz in the King Fahd University of Petroleum & Minerals (KFUPM), Saudi Arabia. They showed that most of the power peaks persisted throughout the time span of spectrogram and that the total amount of power across the 1–2700 MHz band was −14 dBm. Several measurements were conducted in Boston (USA) and analyzed later in [119], where the authors indicated locations and associated RF bands that can point towards the practicality of ambient RF EH. The GSM-850 band was shown to be the most favorable 45% of the time while providing around 3–400 μW in different measurement points, followed by LTE-730 (30%), LTE-740 (16%), GSM-1900 (5%) and DTV (3%). Throughout measurement campaigns in two different urban cities (in France and the

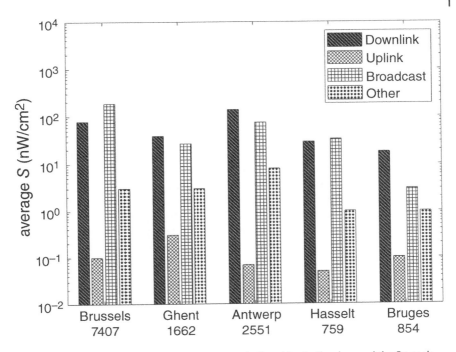

Figure 3.4 Average statistics of S measurements in five cities in Flanders and the Brussels Capital Region in 2017–2019 [201]. The number below the city name indicates the population density in # people/km^2 as retrieved from www.vlaanderen.be on 31/01/2019.

Netherlands) in 2016–2017, the authors of [200] characterized the RF-EMF radiation induced by LTE macro and small-cell BSs. They showed that measurements associated with small cells were usually higher than those associated with macro cells.

In general, the number of publicly available results from RF measurement campaigns is rather limited, especially in recent years. This, and the fact that results from RF measuring campaigns may quickly become obsolete as communication networks tend to densify while new technologies are periodically introduced, suggests the need for many more, and continuous, RF measurement campaign efforts in the coming years.

3.3 Energy Arrival Modeling

EH models play vital roles in 1. network deployment planning, 2. designing efficient energy scheduling, and 3. evaluating the performance of RF–EH-enabled networks. The models for energy arrival/availability in the literature are primarily divided into two classes: deterministic and stochastic models

[179]. However, deterministic models are just suitable for applications with predictable and/or slowly-varying energy arrivals, which is not typically the case in ambient RF–EH scenarios, hence we next focus on stochastic models.

3.3.1 Based on Arbitrary Distributions

Energy arrival is modeled through a sequence of instantaneous realizations of a certain random variable, such as in [205, 206]. For instance, in [205] the authors model ambient EH in mobile ad-hoc networks using arbitrary independently and identically distributed (i.i.d.) energy arrivals, while a chi-squared distribution is used for drawing numerical results. Other distributions commonly used are Bernoulli, Uniform, Poisson and Exponential [179]. This kind of model based on arbitrary distributions is extendable to any ambient energy source, which is also its main drawback for modeling ambient RF–EH. Finally, time-uncorrelated random realizations are usually assumed in order to gain in analytical tractability, although time-correlated models fit better in ambient RF–EH scenarios.

3.3.2 Based on Stochastic Geometry

The amount of harvested energy from RF signals largely depends on two crucial factors: transmit power of surrounded transmitters and the corresponding energy channels (including path loss, shadowing and small-scale fading). Energy arrival models based on stochastic geometry are capable of statistically representing these two factors, thus, they are very attractive for assessing the performance of ambient RF–EH scenarios. Two point processes are prominent to model the deployment of non-dedicated energy transmitters (which may belong to one or several processes). These nodes are spatially and independently distributed over an infinite region in $\mathbb{R}^d$ (usually $\mathbb{R}^2$). These two point processes are:

- Poisson point process (PPP) [207], where the number of RF transmitters in a certain sub-region has a Poisson distribution with mean determined by the specific sub-region intensity measure[5]. For instance, EH from ambient RF downlink signals from a K-tier cellular system and supporting cellular uplink transmissions is investigated in [208], while in [209, 210], the ambient powering signals come from RF transmitters belonging to a single PPP and for supporting device-to-device (D2D) communications. Moreover, a stochastic geometry study of ambient RF–EH considering the non-linearities of the EH process was recently conducted in [211].
- β–Ginibre point process (β–GPP) [212], where there is some dependence on the spatial deployment of the non-dedicated energy transmitters, e.g., as

---

5 In the case of an homogeneous PPP (HPPP), the number of RF transmitters in a certain sub-region has a Poisson distribution with mean proportional to its area.

in [213, 214]. The dependence becomes stronger as β approaches 1 (GPP)[6], while when β = 0, one ends up with the traditional PPP. Therefore, the β–Ginibre point process is a more general model. Note that β–GPP may model more realistic network deployments for which node locations are typically spatially correlated, i.e., there exists repulsion (or attraction) between nodes.

In general, the state-of-the-art literature on stochastic geometry-based ambient energy arrival still has to consider broadband (multi-band or tunable-band) operation as required for efficient ambient RF–EH. Also, more fine-grained information about the capability of a certain region to support ambient RF–EH requires further stochastic studies. In that sense, the novel meta distribution metric, as considered in [210], needs to be more thoroughly investigated since it provides answers to key questions such as *what percentage of the area can support ambient RF–EH-enabled IoT deployments?* We delve into these issues in the next section through a comprehensive performance analysis of ambient RF–EH.

3.4 A Stochastic Geometry-based Study

Using tools from stochastic geometry, we next provide a tractable analytical framework for statistical feasibility analysis of ambient RF–EH.

3.4.1 System Model and Assumptions

Consider the coexistence of $\mathcal{K}$ different types of ambient RF energy sources in $\mathbb{R}^2$. The RF transmitters associated with each $k \in \mathcal{K}$ are spatially distributed as an HPPP denoted as Φ_k with density λ_k. Figure 3.5 illustrates an example deployment. We assume that each RF source $k \in \mathcal{K}$ is associated with a certain transmit power P_k such that the power received from transmitters k at a certain reference point (placed at the origin) is given by

$$p_k = P_k \varphi_k \sum_{i \in \Phi_k} g_{i,k} ||x_{i,k}||^{-\alpha_k}, \qquad \forall k \in \mathcal{K}, \tag{3.1}$$

where $x_{i,k}$ denotes the coordinates of the ith RF transmitter of type k, α_k is the path-loss exponent of its energy transmission channel, and $g_{i,k}$ denotes the small-scale fading power gain coefficient. We assume channels subject to quasi-static Rayleigh fading, thus, $g_{i,k} \sim \text{Exp}(1)$, $\forall k \in \mathcal{K}, i \in \Phi_k$.

---

6 A realization of β–GPP can be obtained from a realization of a GPP. A thinning is performed on the parent GPP such that each point is kept independent from each other with a probability β. A re-scaling with parameter $\sqrt{\beta}$ is performed to preserve the intensity [212].

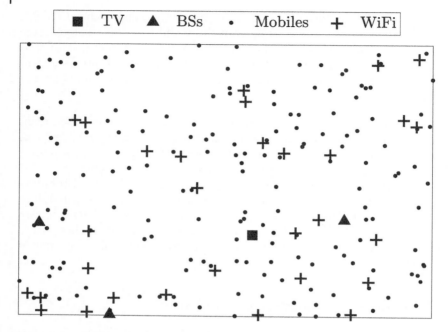

Figure 3.5 Snapshot of the system model in a given area with $|\mathcal{K}| = 4$.

Moreover, φ_k represents the free space path loss at 1 m reference distance, i.e.,

$$\varphi_k = \left(\frac{c}{4\pi f_k}\right)^2, \tag{3.2}$$

where f_k is the nominal operative frequency of RF transmitters corresponding to source k, and c is the speed of light. Finally, we consider normalized time such that power and energy terms are used indistinctly hereinafter.

Single-band Harvesters. Assuming a single-band harvester (top of Figure 3.2) configured to harvest energy just from k−type RF transmitters, and a linear EH model, the harvested power is given by

$$p_{h,k} = \eta_k p_k, \tag{3.3}$$

where p_k is given in (3.1), and $\eta \in (0,1)$ denotes the PCE.

Multi-band Harvesters. Meanwhile, assuming a multi-band harvester (middle of Figure 3.2) configured to harvest energy from a set $\tilde{\mathcal{K}}$ of different RF sources, the harvested power is given by

$$p_{h,\tilde{\mathcal{K}}} = \sum_{k \in \tilde{\mathcal{K}}} \eta_{k,\tilde{\mathcal{K}}} p_k, \qquad \tilde{\mathcal{K}} \subseteq \mathcal{K}. \tag{3.4}$$

Note that we make distinction between η_k in (3.1) and $\eta_{k,\tilde{\mathcal{K}}}$ in (3.4) since in general a multi-band harvester has limited PCE per band such that $\eta_{k,\tilde{\mathcal{K}}} < \eta_k$, $\forall k$.

Tunable-band Harvesters. Finally, herein we assume a tunable-band harvester (bottom of Figure 3.2) configured to dynamically tune the best frequency band among those in $\tilde{\mathcal{K}}$. The best RF source is one that provides the largest average harvested power[7]. Herein, the instantaneous channel fluctuations do not influence the switching/tuning to keep the hardware design simple and energy efficient. Now, the power harvested with a tunable-band device can be written as

$$p_{h,\tilde{\mathcal{K}}}^s = \eta_{k',\tilde{\mathcal{K}}} p_{k'}, \qquad \tilde{\mathcal{K}} \subseteq \mathcal{K}, \tag{3.5}$$

where

$$
\begin{aligned}
k' &= \arg \max_{k \in \tilde{\mathcal{K}}} \eta_{k,\tilde{\mathcal{K}}} \mathbb{E}[p_k | \Phi_k] \\
&= \arg \max_{k \in \tilde{\mathcal{K}}} \eta_{k,\tilde{\mathcal{K}}} P_k \varphi_k \sum_{i \in \Phi_k} ||x_{i,k}||^{-\alpha_k},
\end{aligned} \tag{3.6}
$$

and the last line comes from averaging (3.1) over $g_{i,k}$ which is a unit-mean random variable.

3.4.2 Energy Coverage Probability

The energy coverage probability given a threshold ξ is defined as

$$\mathcal{P}_{p_h}(\xi) = \mathbb{P}[p_h \geq \xi] = 1 - F_{p_h}(\xi), \tag{3.7}$$

where $p_h \in \{p_{h,k}, p_{h,\tilde{\mathcal{K}}}, p_{h,\tilde{\mathcal{K}}}^s\}$, characterizes the percentage of the area where the EH devices harvest more than ξ units of energy, or alternatively, the chances of any randomly selected device to harvest more than ξ units of energy [213, 214]. Note that ξ represents the harvesting power required by the EH devices to operate uninterruptedly, thus, such energy coverage probability is the complement of the energy outage probability. This performance metric is particularly useful in setups where the EH devices lack batteries for energy storage.

---

7 This implies switching over all the frequency bands periodically to evaluate their average performance. In a different implementation, the harvester may stay tuned to a certain frequency band and switch to another one when the performance of the former deteriorates significantly, which would reduce the number of switchings.

The Laplace transform of p_h may be useful for finding $f_{p_h}(x)$ and $F_{p_h}(x)$ via the relations

$$f_{p_h}(x) = \mathcal{L}^{-1}\left(\mathcal{L}_{p_h}(s)\right)\Big|_x, \tag{3.8a}$$

$$F_{p_h}(x) = \mathcal{L}^{-1}\left(\frac{1}{s}\mathcal{L}_{p_h}(s)\right)\Big|_x. \tag{3.8b}$$

In fact, the Laplace transform of the RF power sensed by a certain EH device is analogous to the Laplace transform of the interference experienced by a typical receiver in a wireless communication setup, which has been thoroughly characterized in the literature, [207, 215]. As shown next, such results can be easily extended for the case of the power harvested by single-band and multi-band harvesters under constant PCE.

Theorem 3.1 (Laplace transform of the power harvested by single-band devices)

$$\mathcal{L}_{p_{h,k}}(s) = \exp\left(-\delta(k,\eta_k)s^{2/\alpha_k}\right), \tag{3.9}$$

where $\delta(n,m) = \frac{2\pi^2\lambda_n}{\alpha_n \sin(2\pi/\alpha_n)}\left(mP_n\varphi_n\right)^{2/\alpha_n}$.

Proof. The proof is based on previous results in [207, 215] characterizing the Laplace transform of the interference in HPPP wireless communication networks, and it is included here for completeness.

Using (3.3) and (3.1), the Laplace transform of the power harvested by single-band devices can be obtained as follows

$$
\begin{aligned}
\mathcal{L}_{p_{h,k}}(s) &= \mathbb{E}\left[e^{-sp_{h,k}}\right] = \mathbb{E}\left[\exp\left(-s\eta_k P_k\varphi_k\sum_{i\in\Phi_k}g_{i,k}||x_{i,k}||^{-\alpha_k}\right)\right]\\
&\overset{(a)}{=} \mathbb{E}\left[\prod_{i\in\Phi_k}\exp\left(-s\eta_k P_k\varphi_k g_{i,k}||x_{i,k}||^{-\alpha_k}\right)\right]\\
&\overset{(b)}{=} \mathbb{E}_{\Phi_k}\left[\prod_{i\in\Phi_k}\mathbb{E}_{g_{i,k}}\left[\exp\left(-s\eta_k P_k\varphi_k g_{i,k}||x_{i,k}||^{-\alpha_k}\right)\right]\right]\\
&\overset{(c)}{=} \exp\left(-2\pi\lambda_k\mathbb{E}_{g_{i,k}}\left[\int_0^\infty\left(1-\exp\left(-s\eta_k P_k\varphi_k g_{i,k}y^{-\alpha_k}\right)\right)y\,dy\right]\right)\\
&\overset{(d)}{=} \exp\left(-\frac{2\pi\lambda_k}{\alpha_k}\mathbb{E}_{g_{i,k}}\left[\int_0^\infty\left(1-\exp\left(-s\eta_k P_k\varphi_k g_{i,k}/z\right)\right)z^{2/\alpha_k-1}dz\right]\right)\\
&\overset{(e)}{=} \exp\left(-\pi\lambda_k\mathbb{E}_{g_{i,k}}\left[g_{i,k}^{2/\alpha_k}\right]\Gamma\left(1-\frac{2}{\alpha_k}\right)\left(s\eta_k P_k\varphi_k\right)^{2/\alpha_k}\right), \tag{3.10}
\end{aligned}
$$

where (a) comes from the property $e^{\sum n_i} = \prod e^{n_i}$, (b) comes from exploiting the statistical independence of the HPPP and the channels' fading, (c) follows from the probability generating functional (PGFL) of the HPPP, while (d) from the change of variables $y^{-\alpha_k} \to 1/z$, and (e) from noticing that the argument of the exponential function in (d) resembles the definition of the ith moment of a random variable as in $\mathbb{E}[V^i] = \int iv^{i-1}(1 - F_V(v))dv$, with $i = 2/\alpha_k$ and $V = W^{-1} \sim \mathrm{Exp}\left(\frac{1}{s\eta_k P_k \varphi_k g_{i,k}}\right)$. Finally, (3.9) is attained by using $\mathbb{E}[g_{i,k}^{2/\alpha_k}] = \Gamma(1 + 2/\alpha_k)$ and $\Gamma(1 + 2/\alpha_k)\Gamma(1 - 2/\alpha_k) = \frac{2\pi}{\alpha_k \sin(2\pi/\alpha_k)}$. $\qquad\square$

Note that (3.9) has the form of a scaled stretched exponential function, or Kohlrausch-Williams-Watts (KWW) function [216], for which its inverse Laplace transform, thus, the probability density function (PDF) of $p_{h,k}$, has been derived in closed-form for several integer values of α_k, $\alpha_k \in \{3,4,5,6\}$ [215, Table I]. For simplicity, we focus on $\alpha \in \{4,6\}$, and present in Table 3.5 not only the PDF of a random variable with Laplace transform as given in (3.9) for such α_ks, but also the associated cumulative density functions (CDFs). Since the path-loss exponents are usually smaller than 4 and almost surely smaller than 6 in practice, the performance results associated with such values represent lower bounds of those attained in practice. Meanwhile, for other non-integer values of α, the inverse Laplace transform of $\mathcal{L}_X(s) = \exp(-\delta s^{2/\alpha})$ must be evaluated numerically. By using the complex inversion integral formula for Laplace transforms with $s = 0$ as a branch point of the integrand and the proper Bromwich contour that does not contain this branch point, we attain

$$F_X(x) = 1 - \int_0^\infty \frac{1}{\pi u} e^{-ux - \delta\cos(2\pi/\alpha)u^{2/\alpha}} \sin\left(\delta\sin(2\pi/\alpha)u^{2/\alpha}\right)du, \quad (3.11)$$

$$f_X(x) = \frac{d}{dx}F_X(x) = \frac{1}{\pi}\int_0^\infty e^{-ux - \delta\cos(2\pi/\alpha)u^{2/\alpha}} \sin\left(\delta\sin(2\pi/\alpha)u^{2/\alpha}\right)du. \quad (3.12)$$

In the case of using multi-band harvesters, and based on (3.4) and (3.9), the Laplace transform of $p_{h,\tilde{\mathcal{K}}}$ can be written as

$$\mathcal{L}_{p_{h,\tilde{\mathcal{K}}}}(s) = \prod_{k\in\tilde{\mathcal{K}}} \mathcal{L}_{p_{h,k}}(s)\Big|_{\eta_k \leftarrow \eta_{k,\tilde{\mathcal{K}}}} = \exp\left(-\sum_{k\in\tilde{\mathcal{K}}} \delta(k, \eta_{k,\tilde{\mathcal{K}}})s^{2/\alpha_k}\right), \quad (3.13)$$

Table 3.5 PDF and CDF of X given $\mathcal{L}_X(s) = \exp(-\delta s^{2/\alpha})$.

α	$f_X(x)$	$F_X(x)$
4	$\dfrac{\delta}{2\sqrt{\pi}x^{\frac{3}{2}}}\exp\left(-\dfrac{\delta^2}{4x}\right)$	$1 - \mathrm{erf}\left(\dfrac{\delta}{2\sqrt{x}}\right)$
6	$\dfrac{(\delta/x)^{\frac{3}{2}}}{3\pi}K_{\frac{1}{3}}\left(\dfrac{2\delta^{\frac{3}{2}}}{3\sqrt{3x}}\right)$	$1 - \dfrac{\delta\sqrt{3}\Gamma\left(1-\frac{2}{3}\right)}{2\pi\sqrt[3]{x}}{}_1F_2\left(\frac{1}{3}; \frac{2}{3}, \frac{4}{3}; \frac{\delta^3}{27x}\right) + \dfrac{\delta^2}{6\Gamma(\frac{4}{3})x^{\frac{2}{3}}}{}_1F_2\left(\frac{2}{3}; \frac{4}{3}, \frac{5}{3}; \frac{\delta^3}{27x}\right)$

which can be easily inverted to obtain the PDF and CDF of $p_{h,\tilde{\mathcal{K}}}$ if we consider that all the links are subject to an equal path-loss exponent, i.e., $\alpha_k = \alpha \, \forall k \in \tilde{\mathcal{K}}$. Then, by adopting $\delta \leftarrow \sum_{k \in \tilde{\mathcal{K}}} \delta(k, \eta_{k,\tilde{\mathcal{K}}})$ we can re-use the results in Table 3.5 and (3.11).

The scenario where harvesters are equipped with a tunable-band circuit is more cumbersome to analyze as an optimization procedure is involved. In this case, and according to (3.5) and (3.6), the energy coverage probability can be lower-bounded by

$$\mathcal{P}_{p_{h,\tilde{\mathcal{K}}}^s}(\xi) \geq 1 - F_{p_{h,k^\circ}}(\xi)\big|_{\eta_k \leftarrow \eta_{k^\circ,\tilde{\mathcal{K}}}}, \qquad k^\circ = \arg\max_k \mathbb{P}[K' = k], \qquad (3.14)$$

where $\mathbb{P}[K' = k]$ denotes the probability of tuning the kth band, which according to (3.6) can be obtained as follows

$$\mathbb{P}[K' = k] = \mathbb{P}\big[A_k > A_l, \forall l \in \tilde{\mathcal{K}} \backslash k\big] = \int_0^\infty f_{A_k}(x) \prod_{l \in \tilde{\mathcal{K}} \backslash k} F_{A_l}(x) \mathrm{d}x, \qquad (3.15)$$

where

$$A_k = \eta_{k,\tilde{\mathcal{K}}} \mathbb{E}[p_k | \Phi_k] = \eta_{k,\tilde{\mathcal{K}}} P_k \varphi_k \sum_{i \in \Phi_k} ||x_{i,k}||^{-\alpha_k}. \qquad (3.16)$$

Corollary 3.1 (CDF and PDF of A_k) The CDF and PDF of A_k can be evaluated using (3.11), (3.12) and the results in Table 3.5 with

$$\delta = \pi \lambda_k \Gamma(1 - 2/\alpha_k)(\eta_{k,\tilde{\mathcal{K}}} P_k \varphi_k)^{2/\alpha_k}. \qquad (3.17)$$

Proof. Note that A_k just differs from $p_{h,k}$ in the PCE value and in that $g \sim \mathrm{Exp}(1)$ for the latter, while $g = 1$ (no fading) for the former. Therefore, the Laplace transform of A_k is still given by (3.10) but setting $\mathbb{E}[g_{i,k}^{2/\alpha}] \to 1$ and $\eta_k \to \eta_{k,\tilde{\mathcal{K}}}$. Finally, the CDF and PDF of random variables with Laplace transform of the form (3.10) is given by (3.11) and (3.12), respectively, and in Table 3.5 for $\alpha \in \{4, 6\}$. $\qquad \square$

3.4.3 Average Harvested Energy

Next, we characterize the average energy harvested by a certain EH device in the network, i.e., $\mathbb{E}[p_h], p_h \in \{p_{h,k}, p_{h,\tilde{\mathcal{K}}}, p_{h,\tilde{\mathcal{K}}}^s\}$. Note that according to the adopted path-loss and constant PCE model, $\mathbb{E}[p_h]$ would be infinite since the receive RF power, thus, the harvested power tends to infinity when the RF transmitter is sufficiently close to the harvester, $||x_{i,k}|| \ll 1$ for some i, k. This is obviously a modeling artifact since no receiver ever gets more power than was transmitted. To overcome this, more accurate path-loss models, such as $\min(1, ||x||^{-\alpha})$ or $(1 + ||x||)^{-\alpha}$ [207], and/or non-linear EH models accounting for the saturation

Table 3.6 Average power harvested by single-band devices.

a	$\mathbb{E}[p_{h,k}]$
4	$\delta\sqrt{\dfrac{\rho}{\pi}}\,e^{-\frac{\delta^2}{4\rho}} + \left(\dfrac{\delta^2}{2}+\rho\right)\mathrm{erf}\left(\dfrac{\delta}{2\sqrt{\rho}}\right) - \dfrac{\delta^2}{2}$
6	$\dfrac{\delta^2}{6}\left[\delta + \dfrac{\sqrt{3}\rho^{\frac{1}{3}}}{2\pi}\left(\displaystyle\sum_{\iota\in\{-1,1\}}\left(\dfrac{3\rho^{\frac{1}{3}}}{\delta}\right)^{\frac{\iota}{2}}\Gamma\left(\dfrac{\iota}{3}\right)\left(2\,_1F_2\left(\dfrac{\iota}{3};\dfrac{2}{3},\dfrac{4}{3};\dfrac{\delta^3}{27\rho}\right)+_1F_2\left(-\dfrac{2\iota}{3};\dfrac{5}{6}-\dfrac{\iota}{2},\dfrac{7}{6}-\dfrac{\iota}{2};\dfrac{\delta^3}{27\rho}\right)\right)\right)\right]$

phenomenon, may be used. This was not an issue in the previous section for computing the energy coverage probability since the PDF and CDF of the harvested power do exist, and the focus is on the range of small values of harvested power. For simplicity, herein we adopt the $\min(1,\|x\|^{-\alpha})$ model, then, the average harvested power using single-band harvesters is given by

$$\mathbb{E}[p_{h,k}] = \int_0^\rho x f_{p_{h,k}}(x)\mathrm{d}x + \rho(1 - F_{p_{h,k}}(\rho)), \tag{3.18}$$

where $\rho = \eta_k P_k \varphi_k$ represents an upper bound on the average harvested energy given the new path-loss model, and $f_{p_{h,k}}(x)$ is given in (3.12), and detailed for $\alpha \in \{4,6\}$ in Table 3.5. In fact, by using the results in Table 3.5, closed-form expressions for the average harvested energy using single-band harvesters can be obtained as given in Table 3.6.

According to (3.4) and (3.3), in the case of multi-band harvesters, we have that

$$\mathbb{E}[p_{h,\bar{\mathcal{K}}}] = \sum_{k\in\bar{\mathcal{K}}}\mathbb{E}[p_{h,k}]\big|_{\eta_k \leftarrow \eta_{k,\bar{\mathcal{K}}}}, \tag{3.19}$$

while for tunable-band harvesters, and according to (3.5) and (3.6), a lower-bound for the average harvested energy is given by

$$\mathbb{E}[p^s_{h,\bar{\mathcal{K}}}] \geq \mathbb{E}[p_{h,k^\circ}]\big|_{\eta_k \leftarrow \eta_{k^\circ,\bar{\mathcal{K}}}}, \tag{3.20}$$

where k° is defined in (3.14). Note that $\mathbb{P}[K' = k^\circ] \geq \frac{1}{|\bar{\mathcal{K}}|}$, and the bounds in (3.14) and (3.20) become tighter as this probability increases.

3.4.4 Meta-distribution of Harvested Energy

The meta distribution was formally defined in [217] to provide a much sharper version of the "signal-to-interference ratio (SIR) performance" than that merely considered at the typical link through spatial averaging in two basic Poisson network models. Thereafter, it was analyzed in a similar context for many other communication scenarios, while more recently, in [210], the meta distribution of the harvested energy in wirelessly-powered D2D networks was also proposed and analyzed.

The meta distribution of the harvested energy in ambient RF–EH scenarios, as a fine-grained performance metric, is key for conducting feasibility studies. Concretely, the meta distribution of the harvested energy is the distribution of the energy outage probability conditioned on the locations of the RF sources, defined as

$$\tilde{F}_{\tilde{\mathcal{P}}_{p_h}(\xi)}(\varepsilon) = \mathbb{P}[\tilde{\mathcal{P}}_{p_h}(\xi) > \varepsilon], \qquad \xi \in \mathbb{R}^+, \qquad \varepsilon \in [0,1], \tag{3.21}$$

where $\tilde{\mathcal{P}}_{p_h}(\xi)$ is a random variable given as

$$\tilde{\mathcal{P}}_{p_h}(\xi) = \mathbb{P}[p_h > \xi| \cup_{k \in \mathcal{K}} \Phi_k], \tag{3.22}$$

and $p_h \in \{p_{h,k}, p_{h,\bar{\mathcal{K}}}, p^s_{h,\bar{\mathcal{K}}}\}$ according to the EH circuit topology. Due to the ergodicity of the point processes, the meta distribution can be interpreted as the fraction of devices in each realization of the point processes that have harvested energy above ξ with probability at least ε. The standard energy outage probability is the mean of $\tilde{\mathcal{P}}_{p_h}(\xi)$, obtained by integrating the meta distribution (3.21) over $\varepsilon \in [0,1]$.

An analytical study of the meta distribution, either of the SIR or harvested energy, is generally cumbersome, and the scenario discussed here is not the exception. The most common and tractable approach is based on matching the first two or three moments of the meta-distribution to those of a generalized beta-distribution [210, 217, 218]. However, herein we keep our discussions simple, and just evaluate numerically this metric to illustrate some key features and discuss some performance insights.

3.4.5 Numerical Results

Some numerical results are provided and discussed in this section. The main symbols and parameters are summarized in Table 3.7 together with the default values in the simulations.

Figure 3.7 illustrates the EH performance in terms of energy coverage probability and average harvested energy when using different EH circuit architectures. As for the energy coverage probability, the most favorable frequency band results to be LTE-800 (BTB) given the adopted network specifications. By harvesting energy in this band, the coverage probability is around 5%, 0.4% and 0.004% for energy thresholds of 10 nW, 1 μW and 100 μW, respectively. Obviously, as devices demand more energy to operate (larger energy threshold ξ), the energy coverage probability decreases, which translates to a smaller set of locations and/or time intervals where ambient RF–EH is feasible. The energy coverage probability is considerably reduced when harvesting from DTV,

Table 3.7 Default values for system parameters.

Symbol	Description	Default value
$\mathcal{K}$	set of ambient RF sources	{DTV, LTE-800 (MTB), LTE-800 (BTB), WiFi}
$\tilde{\mathcal{K}}$	set of ambient RF sources for multi-/tunable-band EH	$= \mathcal{K}$
P_k	transmit power of the kth RF source	{1000, 0.25, 50, 0.1} W
f_k	nominal frequency of the kth RF source	{600, 850, 800, 2440} MHz
λ_k	density of the kth RF sources HPPP	$\{10^{-6.5}, 10^{-4}, 10^{-5}, 10^{-3.5}\}$ nodes/m^2
η_k	PCE for a single-band harvester tuned to the kth RF sources	0.6
$\eta_{k,\tilde{\mathcal{K}}}$	PCE in the kth RF band of a multi-/tunable-band harvester	$0.7\eta_k$
α_k	path-loss exponent of transmissions from the kth RF sources	4, $\forall k$
ξ	EH threshold	10 nW

Note that $|\mathcal{K}| = |\tilde{\mathcal{K}}| = 4$ as in Figure 3.5, and the index of the value in curly brackets specifies the corresponding RF source, e.g., LTE-800 (BTB)$\rightarrow k = 3$.

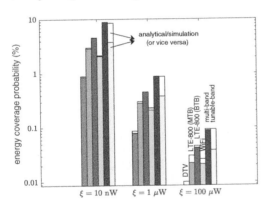

Average harvested energy (dBm)		
Harvester	Analytical	Simulation
DTV	−27.8	−21.1
LTE-800 (MTB)	−41.8	−40.6
LTE-800 (BTB)	−28.3	−27.8
WiFi	−50.0	−49.1
multi-band	−26.4	−25.6
tunable-band	−32.4	−25.6

Figure 3.6 Energy coverage probability for $\xi \in \{10$ nW, 1 μW, 100 μW$\}$ and average harvested energy (dBm). The performance is evaluated for different EH circuit architectures.

LTE-800 (MTB) and WiFi bands. It is worth mentioning that for a specific regional area, using LTE-800 (MTB) or WiFi bands is counterproductive in the sense that their associated RF transmitters are often mobile and with traffic-dependent activation. LTE-800 (BTB) is less sensitive to this, although night periods may still be problematic; while for DTV this is often not an issue. Note that in terms of average harvested energy, DTV seems even to outperform LTE-800 (BTB) mainly because of the relatively high power delivery in the vicinity

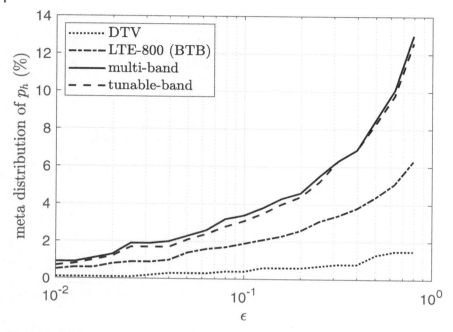

Figure 3.7 Meta distribution of the harvested energy. A single-band (LTE-800 (BTB) and DTV), multi-band, and tunable-band harvesters were considered. The meta distribution performance results in the case of the single-band LTE-800 (MTB) and WiFi harvesters are similar to that of the single-band LTE-800 (BTB) and are not illustrated here.

of the TV towers, and this may be taken conservatively since the EH saturation has not been considered in this analysis. The adoption of more efficient EH circuit architectures such as multi-band or tunable-band harvesters relaxes all the above critical issues. In such cases, the ambient RF–EH feasibility increases, both in terms of energy coverage probability, reaching values around 9%, 0.8% and 0.13% for $\xi = 10$ nW, 1 μW, 100 μW, respectively, and average EH performance, allowing -25.6 dBm of energy to be harvested on average. Finally, observe that the analytical and simulation results significantly match.

The performance in terms of the meta distribution of the harvested energy is depicted in Figure 3.7. Herein, the meta distribution characterizes the portion of the network area where a given IoT device is able to harvest enough energy to support its operation at least $100 \times (1 - \varepsilon)\%$ of the time. By deploying single-band harvesters, one may notice that ambient RF EH is feasible 90% of the time in 0.4% (DTV) and 1.9% (LTE-800 (BTB)) of the area. Meanwhile, by using more efficient EH circuit architectures such as multi-band or tunable-band harvesters, the area extends to 3.4% and 3.1%, respectively. Obviously, the harsher the QoS requirement, the smaller the area where ambient RF EH is feasible. For instance, if the time requirement is set to 99% ($\varepsilon = 10^{-2}$), the area percentages reduce to 0.1% (DTV), 0.5% (LTE-800 (BTB)), 0.9% (multi-band) and 0.7% (tunable-band).

3.5 Final Considerations

In this chapter, we investigated the ambient RF EH technology. We provided a classification according to the nature of the ambient RF source, say, static or dynamic, while showing key typical examples of each. The possibility of using hybrid, EH–power grid, implementations for meeting more demanding energy requirements was discussed in the context of future wireless systems. Moreover, we surveyed three key energy usage protocols, HU, HSU and HUS, and exemplified their relative performance. In this regard, we showed that the often adopted HU protocol performs the worst since no energy saving is allowed, while the superiority level of HUS and HSU protocols depends on the battery impairments, being more favorable for HUS in general.

Some approaches for enabling efficient ambient RF–EH were discussed: adopting high-gain multi-directional rectennas, deploying large EH surfaces and tunable-band harvesters. Specifically, a high-gain multi-directional rectenna boosts the RF energy at the input of the rectifiers and its adoption becomes attractive/feasible for moderate/large-sized devices. Meanwhile, deploying large harvesting surfaces in building facades, room walls/ceilings, may contribute to recycle significant amounts of ambient energy. The notion of a tunable-band harvester was proposed as an efficient alternative to single-band circuits, which are strictly sub-optimal for ambient RF–EH applications, and multi-band circuits, which often require more complex hardware designs. Moreover, we discussed relatively recent ambient RF measurement campaigns, and motivated the need for stochastic models based on arbitrary distributions and stochastic geometry for energy arrival characterization.

We also presented in this chapter a stochastic geometry-based study where EH IoT devices were deployed in a large area overlapped with different RF sources. For numerical evaluation, the RF sources were chosen to be TV towers, LTE BSs, LTE mobiles and WiFi transmitters, and were modeled as four independent HPPPs. We discussed the system performance in terms of energy coverage probability, average harvested energy and the meta distribution of the harvested energy in case of single-band, multi-band and tunable-band harvesters. The considerable performance improvements brought by multi-band and tunable-band harvesters motivate their adoption in practical setups.

4

Efficient Schemes for WET

4.1 EH from Dedicated WET

In this chapter, we focus on RF–EH from dedicated WET. In Chapter 3, we discussed different ambient RF–EH applications, conversely, as we shall see in this chapter, EH circuits designed to be powered via dedicated WET may not require multi-band operation support. Note that the use of multiple EH antennas is still attractive for boosting the harvesting capability, but contrary to the schemes presented in Figure 3.4, they may be designed for (and tune to) a given frequency band (the specific band in which its RF transmitter operates). However, the use of multiple antennas at the harvesting device may be prohibitive due to form-factor constraints in many IoT applications, for which single-antenna EH devices are preferable[1]. Herein, we focus on the latter, although many of the exposed ideas, methods and algorithms can be easily adapted to fit scenarios with multi-antenna harvesters.

4.2 Energy Beamforming

Consider a PB equipped with M antennas and transferring energy via RF to a set $\mathcal{S} = \{S_1, S_2, \cdots, S_N\}$ of N single-antenna IoT devices located nearby. Quasi-static channels are assumed, with fading remaining constant over a transmission block, and changing from block to block with a certain distribution. The power-normalized channel vector between the PB's antennas and S_i is denoted as $\mathbf{h}_i \in \mathbb{C}^{M \times 1}$, while ϱ_i denotes the average power gain of the channel between the PB and S_i.

---

1 Such an assumption is in line with current technologies, as discussed in Chapter 1.

Wireless RF Energy Transfer in the Massive IoT Era: Towards Sustainable Zero-energy Networks.
First Edition. Onel Alcaraz López and Hirley Alves.
© 2022 by The Institute of Electrical and Electronics Engineers, Inc. Published 2022 by John Wiley & Sons, Inc. Companion website: www.wiley.com/go/Alves/WirelessEnergyTransfer

The PB transmits $K \leq \min(M, N)$ complex signals $\{x_k\}$ such that the received RF signal available at each device S_i is given by

$$y_i = \sqrt{\varrho_i} \mathbf{h}_i^T \sum_{k=1}^K \mathbf{w}_k x_k, \qquad i = 1, \cdots, N, \tag{4.1}$$

where $\mathbf{w}_k \in \mathbb{C}^{M \times 1}$ represents the precoding vector associated with x_k. For simplicity noise is ignored and transmit power is normalized. Signals are assumed independent and normalized such that $\mathbb{E}[x_k^H x_k] = 1$ and $\mathbb{E}[x_k^H x_j] = 0, \forall k \neq j$. As such, the RF energy (normalized by unit time) available at each S_i is given by

$$E_i = \mathbb{E}_x[y_i^H y_i] = \varrho_i \mathbb{E}_x \left[\left(\sum_{k=1}^K \mathbf{h}_i^T \mathbf{w}_k x_k \right)^H \left(\sum_{k=1}^K \mathbf{h}_i^T \mathbf{w}_k x_k \right) \right]$$

$$= \varrho_i \sum_{k=1}^K \left| \mathbf{h}_i^T \mathbf{w}_k \right|^2 \mathbb{E}_x[x_k^H x_k]$$

$$= \varrho_i \sum_{k=1}^K \left| \mathbf{h}_i^T \mathbf{w}_k \right|^2, \tag{4.2}$$

while the energy harvested by S_i,

$$E_i^h = g(E_i), \tag{4.3}$$

depends on the non-linear EH transfer function $g : \mathbb{R}^+ \mapsto \mathbb{R}^+$ as surveyed in Section 2.2.

With the above formulation in place, we can cast different optimization problems. We highlight four canonical problems, which are illustrated below, for scenarios with battery-less EH devices that often demand a memory-less optimization (there is no battery state evolution, and hence, powering strategies taken at some point do not often affect future actions[2]). These problems can be combined and/or enriched with weighting factors to conform other more evolved optimization systems. Herein, we focus on the maximum fairness problem[3] **P2**, while for simplicity but without loss of generality, we set the total transmit power constraint to the unit. Notice that the transmit power constraint of **P2** is convex, while the objective function is not concave, therefore the problem is not convex. However, it can still be optimally solved by rewriting it as a semi-definite programm (SDP) [219] as shown next.

---

2 A counter example is the case where one needs to keep a record of the devices that have been favored in previous transmissions with a good energy transfer process in order to favor others in future instances.

3 In fact, the solution of **P2** matches the solution of **P4** when $\xi_i = \xi, \forall i$, but scaled to satisfy the corresponding constraints.

Memory-less canonical WET optimization problems

Max EH with power constraint

$$\textbf{P1:} \quad \max_{\{\mathbf{w}^{(j)}\},\, \forall j} \quad \sum_{i=1}^{N} E_i^h$$

$$\text{s. t.} \quad \sum_{k=1}^{K} ||\mathbf{w}_k||^2 \le P.$$

Max fairness with power constraint

$$\textbf{P2:} \quad \max_{\{\mathbf{w}^{(j)}\},\, \forall j} \quad \inf_{i=1,\cdots,N} E_i^h$$

$$\text{s. t.} \quad \sum_{k=1}^{K} ||\mathbf{w}_k||^2 \le P.$$

Min power with total EH constraint

$$\textbf{P3:} \quad \min_{\{\mathbf{w}^{(j)}\},\, \forall j} \quad \sum_{k=1}^{K} ||\mathbf{w}_k||^2$$

$$\text{s. t.} \quad \sum_{i=1}^{N} E_i^h \ge \xi.$$

Min power with EH constraints

$$\textbf{P4:} \quad \min_{\{\mathbf{w}^{(j)}\},\, \forall j} \quad \sum_{k=1}^{K} ||\mathbf{w}_k||^2$$

$$\text{s. t.} \quad E_i^h \ge \xi_i, \;\; \forall i.$$

First, notice that because of the non-decreasing behavior of g, one can directly substitute E_i^h by E_i in the objective function of **P2** without affecting the optimization result. Interestingly, this does not hold for **P1** and **P3**. Second, define $\xi \triangleq \inf E_i$, while E_i in (4.2) can be rewritten as

$$E_i = \varrho_i \sum_{k=1}^{K} \left| \mathbf{h}_i^T \mathbf{w}_k \right|^2 \varrho_i \sum_{k=1}^{K} \mathbf{h}_i^H \mathbf{w}_k \mathbf{w}_k^H \mathbf{h}_i$$

$$= \varrho_i \, \text{Tr}(\mathbf{W} \mathbf{H}_i), \tag{4.4}$$

where $\mathbf{W} = \sum_{k=1}^{K} \mathbf{w}_k \mathbf{w}_k^H$, $\mathbf{W} \in \mathbb{C}^{M \times M}$ and $\mathbf{H}_i = \mathbf{h}_i \mathbf{h}_i^H$. Third, notice that $\mathbf{W}$ is a Hermitian matrix with maximum rank $\min(N, M)$ that can be found by solving the SDP:

$$\textbf{P2.1:} \quad \min_{\mathbf{W} \in \mathbb{C}^{M \times M},\, \xi} \quad -\xi \tag{4.5a}$$

$$\text{s. t.} \quad \varrho_i \, \text{Tr}(\mathbf{W} \mathbf{H}_i) \ge \xi, \quad i = 1, \cdots, N \tag{4.5b}$$

$$\text{Tr}(\mathbf{W}) = 1 \tag{4.5c}$$

$$\mathbf{W} \succeq 0. \tag{4.5d}$$

Notice that constraint (4.5c) constitutes the transmit power constraint here. Finally, the beamforming vectors $\{w_k\}$ match the eigenvectors of $\mathbf{W}$ but normalized by their corresponding eigenvalues' square roots such that $\mathrm{Tr}(\mathbf{W}) = 1$. This approach is referred hereinafter as **optimum full-CSI beamforming**.

Interior point methods (IPMs) are usually adopted to efficiently solve SDP problems [220, 221]. In this case, it can be shown that solving **P2.1** requires around $\mathcal{O}(\sqrt{M}\ln(1/\varepsilon))$ iterations, with each iteration requiring at most $\mathcal{O}(M^6 + (N+1)M^2)$ arithmetic operations [221], and where ε represents the solution accuracy attained when the algorithm ends. Consequently, the SDP solution becomes computationally costly as the number of antennas at the PB increases. To achieve complexity reduction, a low-complexity beamforming design was proposed in [85] as an efficient alternative for scenarios where $N \leq M$. Moreover, this algorithm attains near-optimum performance as the channels become more deterministic. This approach is detailed in the next section.

4.2.1 Low-complexity EB Design

Herein, it is assumed that $N \leq M$. Then, the complexity of **P2.1** can be alleviated by adopting [85]

$$\mathbf{w}_k^T = \frac{\mathbf{h}_k^H}{||\mathbf{h}_k||}\sqrt{p_k}, \qquad k = 1, \cdots, K, \tag{4.6}$$

with $K = N$. In doing this, the ith signal transmitted from all PB's antennas arrives at S_i with constructive superposition. In fact, this design is akin to the maximum ratio transmission (MRT) in multiple-input single-output (MISO) communications. Notice that $\mathbf{p} = [p_1, p_2, \cdots, p_K]$, where p_k represents the power budget for x_k such that $\sum_{i=1}^{N} p_i = 1$, and the impact of the signals for other devices on the RF energy at S_i is not considered for the phase's design. Thus, the RF energy available at S_i when the PB adopts (4.6) can be written as

$$E_i = \varrho_i \, \mathrm{Tr}(\mathbf{W}\mathbf{H}_i) = \varrho_i \sum_{k=1}^{N} \left| \frac{\mathbf{h}_k^H \mathbf{h}_i \sqrt{p_k}}{||\mathbf{h}_k||} \right|^2 = \varrho_i \sum_{k=1}^{N} Q_{k,i} p_k, \tag{4.7}$$

where $Q_{k,i} = |\mathbf{h}_k^H \mathbf{h}_i|^2 / ||\mathbf{h}_k||^2$ represents the fraction of power contribution at S_i of the signal meant to S_k, and notice that $Q_{k,k} = \mathbf{w}_k^T \mathbf{h}_k = ||\mathbf{h}_k||^2$. With the above result, **P2.1** can be re-written as a linear program (LP) as

$$
\textbf{P2.2:} \quad \min_{\mathbf{p}, \xi} \quad -\xi \tag{4.8a}
$$

$$
\text{s. t.} \quad \mathbf{BQ}^T\mathbf{p} \succeq \xi \mathbf{1}_{N \times 1} \tag{4.8b}
$$

$$
\mathbf{1}^T\mathbf{p} = 1 \tag{4.8c}
$$

$$
\mathbf{p} \succeq \mathbf{0}, \tag{4.8d}
$$

where $\mathbf{B} = \mathrm{diag}([\varrho_1, \varrho_2, \cdots, \varrho_N])$. The problem is now composed of $N + 1$ linear constraints and variables, and can be solved efficiently. For instance, if IPMs are used, solving **P2.2** will take at most $\mathcal{O}(\sqrt{N+1}\log(1/\varepsilon))$ iterations, each one with at most $\mathcal{O}((N + 1)^3)$ arithmetic operations [221]. This is a considerable complexity reduction compared to the optimum SDP-based solutions described in the previous section. Moreover, the authors in [85] proposed a low-complexity EB (LCEB) algorithm for solving **P2.2**, as shown in Algorithm 1. LCEB constitutes a particularly simple interior-point method implementation based on affine scaling [221].

LCEB is derived as follows. First, **P2.2** is transformed to the form:

$$
\min_{\mathbf{z}} \quad \mu^T\mathbf{z} \tag{4.9}
$$

$$
\text{s. t.} \quad \mathbf{Az} = \mathbf{b},
$$

$$
\mathbf{z} \succeq \mathbf{0},
$$

Algorithm 1 Low-complexity EB (LCEB)

1: **Input:** $\mathbf{Q}$, $\mathbf{B}$, $\delta \in (0, 1)$ and $\varepsilon \in (0, 1)$

2: Set $\mathbf{A} = \begin{bmatrix} \mathbf{1}_{N \times 1} & -\mathbf{BQ}^T & \mathbf{I}_{N \times N} \\ 0 & \mathbf{1}_{1 \times N} & \mathbf{0}_{1 \times N} \end{bmatrix}$, $\mu = [-1, 0, \cdots, 0]^T$

3: Set $\mathbf{p}_0$ using (4.12), $\xi_0 = \min\{E_i|_{\mathbf{p}_0}\}$ and $\nu = E_i|_{\mathbf{p}_0} - \xi_0$

4: Set $\mathbf{z}^{(0)} = [\xi_0, \mathbf{p}_0, \nu]^T$ and $\tau = 0$

5: **repeat**

6: $\quad \mathbf{Z}^{(\tau)} = \mathrm{diag}(\mathbf{z}^{(\tau)})$, $\lambda^{(\tau)} = \left(\mathbf{A}(\mathbf{Z}^{(\tau)})^2\mathbf{A}^T\right)^{-1}\mathbf{A}(\mathbf{Z}^{(\tau)})^2\mu$

7: $\quad \mathbf{r}^{(\tau)} = \mu - \mathbf{A}^T\lambda^{(\tau)}$

8: $\quad \mathbf{z}^{(\tau+1)} = \mathbf{z}^{(\tau)} - \delta(\mathbf{Z}^{(\tau)})^2\mathbf{r}^{(\tau)}/||\mathbf{Z}^{(\tau)}\mathbf{r}^{(\tau)}||$

9: $\quad \tau \leftarrow \tau + 1$

10: **until** $\mathbf{1}^T\mathbf{Z}^{(\tau-1)}\mathbf{r}^{(\tau-1)} < \varepsilon$ and $\mathbf{r}^{(\tau-1)} \succeq \mathbf{0}$

11: **Output:** $\{p_i = z_{i+1}^{(\tau)}\}_{i=1,\cdots,N}$ and τ

---

which is done by stacking constraints (4.8b) and (4.8c) into a single system of equations with μ and $\mathbf{A}$ given in line 2. Note that the variable space $\mathbf{z}$ now groups ξ, $\mathbf{p}$ and the slack vector $\mathbf{v} \succeq 0$ as defined and initialized in line 4, while $\mathbf{b} = [\mathbf{0}_{1 \times N}, 1]^T$. For initialization, $\mathbf{p}$ is computed in line 3, and although it can be any positive vector satisfying the power constraint, it is preferable to set its value according to (4.12) (derived in the proof of Theorem 4.1). Therein, we shall show (4.12) to be optimal as the number of antennas approaches infinity. Thus, it should provide a good initial guess. At the same time, ξ is set to be the minimum RF energy available at the devices when using such power allocation; and $\mathbf{v}$ is the corresponding slack vector. In addition to the initial $\mathbf{z}$, the iteration index τ is also established in line 4. Lines 5–10 constitute the core of the affine scaling method. Specifically, lines 6, 7 are for computing the dual estimates, λ, and reduced costs $\mathbf{r}$, while line 8 is the updating step consisting of an affine scaling with coefficient δ. The algorithm stops, i.e., convergence is declared, when no cost remains negative[4] ($\mathbf{r} \succeq \mathbf{0}$) and the variation in the objective function is already inferior to the tolerance error ε. Then, the power allocation returned by LCEB in line 11 is ε-optimal.

The approach described in this section is referred to as **full-CSI LCEB** hereinafter.

Performance Bounds.

Theorem 4.1 E_i in (4.7) is respectively upper- and lower-bounded by

$$E_i \leq E_{\text{ub}} = \min\{\varrho_i \|\mathbf{h}_i\|^2\} \mathbf{1}^T \mathbf{p} = \min\{\varrho_i \|\mathbf{h}_i\|^2\}, \tag{4.10}$$

$$E_i \geq E_{\text{lb}} = \frac{1}{\sum_{k=1}^N \varrho_k^{-1} \|\mathbf{h}_k\|^{-2}}. \tag{4.11}$$

Proof. The upper-bound in (4.10) comes from taking advantage of the fairness of the problem solution, the total power budget, and the upper bound of $Q_{k,i} \leq Q_{i,i} = \|\mathbf{h}_i\|^2$, which comes from using the Cauchy – Schwarz inequality [220]. Meanwhile, a lower bound can be derived by noticing that all entries of matrix $\mathbf{Q}$ are non-negative. Then, consider the extreme case $\mathbf{Q} = \text{diag}\left(\{\|\mathbf{h}_i\|^2\}_{\forall i}\right)$ under which **P2.2** can be easily solved to obtain

$$p_i^{\text{opt}} = \frac{1/(\varrho_i Q_{i,i})}{\sum_{k=1}^N 1/(\varrho_k Q_{k,k})} = \frac{1}{1 + \sum_{k \neq i} \frac{\|\mathbf{h}_i\|^2}{\|\mathbf{h}_k\|^2} \frac{1}{\varrho_k}}. \tag{4.12}$$

By substituting (4.12) into (4.7), one attains (4.11). □

---

4 In practice, it is usually required to relax this and allow the algorithm to stop even if $\mathbf{r}$ still contains very small negative values. This is to counteract numerical precision errors, hence one should use $\mathbf{r} \succeq -\zeta \mathbf{1}_{N \times 1}$ with small ζ. In Section 4.2.3, $\zeta = 10^{-4}$ is used.

Note that the lower bound in (4.11) is not attainable unless all devices have the same channel, which will not happen in practice, or when the path-loss of a certain device is much larger than that of the others such that its received energy dominates in the beamforming design.

In general, $||\mathbf{h}_i||^2$ increases linearly with M. Hence, both bounds, (4.11) and (4.10), grow linearly and unbounded[5] with M, and consequently it is possible to conclude that the actual E_i, $\forall i$, also grows with M. Additionally, note that $E_{ub} \leq NE_{lb}$ due to the inequality between the harmonic mean and the minimum function (i.e., $\min\{v_i\} \leq N/\sum_{k=1}^{N} v_k^{-1}$). Therefore, the gap between E_{ub} and E_{lb} is limited by the number of EH devices.

Corollary 4.1 *As M grows larger, the LCEB (Algorithm 1) converges faster.*

Proof. According to the complexity analysis of **P2.2** at the beginning of this subsection, the limiting number of required iterations and arithmetic operations does not depend on M. Meanwhile, by increasing M, the LCEB solution gets closer to the initial guess. This is because by increasing M, $\mathbf{Q}$'s off-diagonal elements decrease, and $\mathbf{Q}$ tends to asymptotically ($M \to \infty$) mimic a diagonal matrix (due to the channel-hardening effect), for which (4.12) is the optimal power allocation. Hence, a faster convergence as M increases. $\qquad\qquad\qquad\qquad\qquad\qquad\qquad\qquad\qquad\qquad\qquad\qquad\qquad\square$

4.2.2 CSI-limited Energy Beamforming

So far we have been concerned with solving **P2** assuming the instantaneous CSI vectors are perfectly known at the PB. However, in practice, not only CSI is imperfect but the devices' cooperation is also required for its acquisition, which consumes their harvested energy. In some cases, the EB gains cannot compensate the energy consumed during the CSI acquisition, and as a result, the net harvested energy of devices become negative [222, 223]. The CSI acquisition overhead may be reduced by noticing that farthest users' channels are expected to dominate the beamforming design, and consequently the PB may not assign pilots for CSI acquisition to the closest EH devices [86]. However, this does not help significantly since the farthest (more energy-limited) users would still be required to spend valuable energy resources for channel training.

An attractive solution lies in exploiting statistical knowledge of the channels, so-called partial/statistical CSI instead of instantaneous CSI. This is three-fold advantageous [78, 85]:

---

5 Note that such analytical results do not violate the law of conservation of energy in practice due to the more substantial path-loss as compared to beamforming gain.

- such information varies over a much larger time scale and does not require frequent CSI updates;
- it is learned with limited CSI-acquisition overhead and energy expenditure;
- it is less prone to estimation errors.

To illustrate these points, we focus on the average harvested energy optimization. To do so, let us first decouple the channel coefficients as

$$\mathbf{h}_i = \bar{\mathbf{h}}_i + \hat{\mathbf{h}}_i, \tag{4.13}$$

where $\bar{\mathbf{h}}_i$ is the deterministic component and $\hat{\mathbf{h}}_i$ is the zero-mean random component with covariance $\mathbf{R}_i = \mathbb{E}[\hat{\mathbf{h}}_i \hat{\mathbf{h}}_i^H]$. Then, it can be observed that

$$
\begin{aligned}
\mathbb{E}[E_i] &= \mathbb{E}\left[\varrho_i \operatorname{Tr}(\mathbf{W}\mathbf{H}_i)\right] \\
&\stackrel{(a)}{=} \varrho_i \mathbb{E}\left[\operatorname{Tr}(\mathbf{W}\bar{\mathbf{H}}_i) + \operatorname{Tr}(\mathbf{W}\tilde{\mathbf{H}}_i) + \operatorname{Tr}(\mathbf{W}\hat{\mathbf{H}}_i)\right] \\
&\stackrel{(b)}{=} \varrho_i \operatorname{Tr}\left(\mathbf{W}(\bar{\mathbf{H}}_i + \mathbf{R}_i)\right),
\end{aligned} \tag{4.14}
$$

where (a) comes from

$$
\begin{aligned}
\mathbf{H}_i &= \mathbf{h}_i \mathbf{h}_i^H = (\bar{\mathbf{h}}_i + \hat{\mathbf{h}}_i)(\bar{\mathbf{h}}_i + \hat{\mathbf{h}}_i)^H \\
&= \underbrace{\bar{\mathbf{h}}_i \bar{\mathbf{h}}_i^H}_{\bar{\mathbf{H}}_i} + \underbrace{\bar{\mathbf{h}}_i \hat{\mathbf{h}}_i^H + \hat{\mathbf{h}}_i \bar{\mathbf{h}}_i^H}_{\tilde{\mathbf{H}}_i} + \underbrace{\hat{\mathbf{h}}_i \hat{\mathbf{h}}_i^H}_{\hat{\mathbf{H}}_i},
\end{aligned} \tag{4.15}
$$

and (b) comes from taking the expectation inside the trace, which is a linear operator, and using $\mathbb{E}[\bar{\mathbf{H}}_i] = \bar{\mathbf{H}}_i$, $\mathbb{E}[\tilde{\mathbf{H}}_i] = \mathbf{0}$, and $\mathbb{E}[\hat{\mathbf{H}}_i] = \mathbf{R}_i$. From (4.14), it becomes evident that the optimum, or low-complexity near-optimum, statistical-CSI beamforming, $\{\mathbf{w}_k^{\mathrm{opt}}\}_{\forall k}$, can be obtained by solving **P2.1**, or **P2.2**, respectively, but using $\bar{\mathbf{H}}_i + \mathbf{R}_i$ instead of $\mathbf{H}_i$. Note also that $\mathbb{E}[\mathbf{H}_i] = \bar{E}_i + \hat{E}_i$, where $\bar{E}_i = \varrho_i \operatorname{Tr}(\mathbf{W}\bar{\mathbf{H}}_i) > 0$ and $\hat{E}_i = \varrho_i \operatorname{Tr}(\mathbf{W}\mathbf{R}_i) > 0$ correspond to the average energy associated to the first- and second-order channel statistics, respectively. Herein, we consider only average CSI is available. Thus, only $\{\bar{\mathbf{H}}_i\}$ are assumed to be known. This information is expected to be beneficial since WET channels are typically LOS-dominant due to the short distances, and consequently have strong deterministic components[6]. By using $\bar{\mathbf{H}}_i$ instead of $\mathbf{H}_i$, we refer to **optimum average-CSI** beamforming when employing **P2.1**. While we refer to **average-CSI LCEB** when employing **P2.2**. Both solutions approach their full CSI counterpart when the channels tend to be fully deterministic, such that $\hat{\mathbf{h}}_k \to \mathbf{0}$, $\forall k$.

---

6 CSI acquisition procedures may be further reduced by exploiting information related to the PB's antenna array architecture and the positioning of the devices [87], which influence the LOS channel, and consequently the channel's deterministic component, the most.

In the case of the **average-CSI LCEB**, the performance bounds in Theorem 4.1, and consequently Corollary 4.1, hold by using $\bar{\mathbf{h}}_i$ instead of $\mathbf{h}_i$, and $\bar{E}_i$ instead of E_i. In fact, the corresponding lower bound for $\bar{E}_i$ would serve also as a lower bound for the actual $\mathbb{E}[E_i]$. Meanwhile, as the spatial correlation (positively) increases, the average energy actually delivered may be much larger than that predicted by (4.11), and even reach (or surpass) the upper bound provided in (4.10).

4.2.3 Performance Analysis

Herein, Rician fading channels are considered with different LOS factors $\kappa_i \geq 0$ [224, Ch. 2]. Under such fading, $\bar{\mathbf{h}}_i$ corresponds to the LOS component of the channel, while $\hat{\mathbf{h}}_i$ represents the scattering contribution. They are respectively given by

$$\bar{\mathbf{h}}_i = \sqrt{\frac{\kappa_i}{1 + \kappa_i}} e^{\mathrm{i}(\boldsymbol{\Phi}_i + \varphi_0)}, \tag{4.16}$$

$$\hat{\mathbf{h}}_i \sim \sqrt{\frac{1}{1 + \kappa_i}} \mathcal{CN}(\mathbf{0}, \mathbf{R}), \tag{4.17}$$

by assuming that signals from all antennas experience the same average path-loss. Note that $\boldsymbol{\Phi}_i$ in (4.16) is what S_i observes as the mean phase shift vector among the PB's antenna elements. Assuming a uniform linear array (ULA) at the PB, we have that

$$\boldsymbol{\Phi}_i = -[0, 1, \cdots, (M - 1)]^T \pi \sin \theta_i, \tag{4.18}$$

where θ_i is the azimuth angle of terminal S_i relative to the boresight of the PB's antenna array [160, Ch. 5]. Meanwhile, φ_0 in (4.16) accounts for an initial phase shift, which without loss of generality we conveniently set equal to $\pi/4$ [225] so $\exp(\mathrm{i}\varphi_0) = (1 + \mathrm{i})/\sqrt{2}$ imposes the effect of LOS (constant) component on real and imaginary parts of the scattering (Rayleigh) component $\hat{\mathbf{h}}$. Herein, we further assume uncorrelated channels, thus $\mathbf{R} = \mathbf{I}$.

Insights on the Average RF Energy Available at Each Device. Based on the above assumptions, it holds that $||\bar{\mathbf{h}}_i||^2 = \frac{\kappa_i}{1+\kappa_i}M$, and consequently (4.10) and (4.11) are modified such that

$$\frac{M}{\sum_{k=1}^N \frac{\kappa_k+1}{\varrho_k \kappa_k}} \leq \bar{E}_i \leq M \min\left\{ \frac{\varrho_i \kappa_i}{1 + \kappa_i} \right\}. \tag{4.19}$$

Finally, note that

$$\hat{E}_i = \frac{\varrho_i}{1 + \kappa_i} \operatorname{Tr}(\mathbf{WI}) = \frac{\varrho_i}{1 + \kappa_i} \operatorname{Tr}(\mathbf{W}) = \frac{\varrho_i}{1 + \kappa_i} \tag{4.20}$$

Table 4.1 EB designs analyzed in this section.

EB design	Main assumptions	Solver
optimum full-CSI	instantaneous CSI	SDP global solver
full-CSI LCEB	instantaneous CSI, $N \leq M$	Algorithm 1
optimum average-CSI	average CSI	SDP global solver
average-CSI LCEB	average CSI, $N \leq M$	Algorithm 1

since $\mathbf{R}_i = \frac{1}{1+\kappa_i}\mathbf{I}$ and $\text{Tr}(\mathbf{W}) = 1$. Observe that even under non-LOS conditions, when $\kappa = 0$, the proposed beamforming can provide the same energy level as in a single-antenna PB system. Obviously, such energy will be larger when κ increases and/or under the effect of some positive spatial correlation.

Numerical Results. Next we present numerical results on the performance of previously discussed EB designs, which are summarized in Table 4.1. It is worth highlighting that the analyzed schemes do not assume any fading distribution knowledge, thus they work for arbitrary channels. LCEB algorithm (Algorithm 1) is run with affine scaling coefficient $\delta = 0.9$ and error tolerance $\varepsilon = 10^{-5}$.

The EH devices are assumed to be randomly and uniformly distributed around the PB at distances between 1 and 10 m, in an annulus region of around 311 m^2 of area. A log-distance path-loss model with exponent 2.7 is considered along with a non-distance dependent loss of 16 dB [226], i.e.,

$$\varrho_i = 10^{-1.6} \times d_i^{-2.7}, \tag{4.21}$$

where d_i is the distance between S_i and the PB. Unless stated otherwise, $M = N = 8$, and $\kappa = 10$ dB is set for all Rician channels involved.

Figure 4.1 corroborates that despite its simplicity, the LCEB design offers a performance that approaches that of the optimum SDP-based implementation. In fact, it even outperforms the optimum average-CSI design when using just average CSI and the Rician factor is below 15 dB. As the Rician factor increases, the performance gap between the optimum full-CSI and the two average-CSI schemes diminishes[7]. In addition, Figure 4.1 validates the bounds given in (4.19), and shows that for this particular scenario, the upper (lower) bound is tighter under small (large) Rician factor.

Figure 4.2(a) validates the results in previous sections predicting a nearly linear performance improvement with the number of antennas M at the PB. Notice that as the number of devices increases, the chance of being farther from the PB increases, thus, deteriorating the system performance. Meanwhile,

---

7 This is without considering the power consumed in the CSI acquisition, which would tilt the scale in favor of the average-CSI schemes.

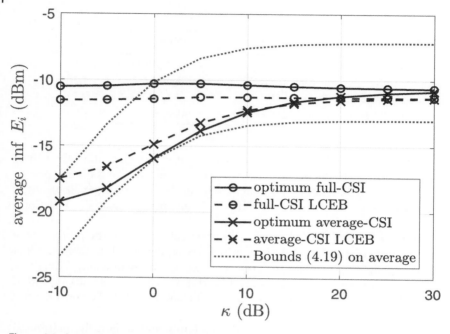

Figure 4.1 Average worst-case RF energy available at the EH devices.

the low-complexity feature of the LCEB design is illustrated in Figure 4.2(b). As expected, as the number of devices increases, more iterations are required. Different from the SDP-based approaches, increasing the number of antennas is shown to be beneficial when using LCEB as predicted in Corollary 4.1. All in all, the fast convergence feature of LCEB contrasts with the SDP-based implementations (both full-CSI and average-CSI) which not only require considerably more iterations, but also, each iteration takes more time as M increases.

On the other hand, note that the ULA angular orientation directly impacts on θ_i, and consequently on ϕ_i, h_i and Q, influencing the system performance. As in [85], assume the PB can adjust its orientation by rotating the array by θ_0 radians such that $\theta_i \leftarrow \theta_i + \theta_0$. This may be possible in static setups, where the task is committed to the technician/user, or in slow-varying environments, where the PB itself is equipped, for instance, with a rotary-motor as in [227]. The performance of the average-CSI EB designs as a function of such rotation angle is shown in Figure 4.3 for three different setups. Notice that the antenna array orientation and/or devices' angular position play a major role on the system performance. Intuitively, it would be desirable that the ULA is geared towards the farthest user(s) to counteract the most adverse path-loss(es); however, this is not completely true and the impact of other devices' position

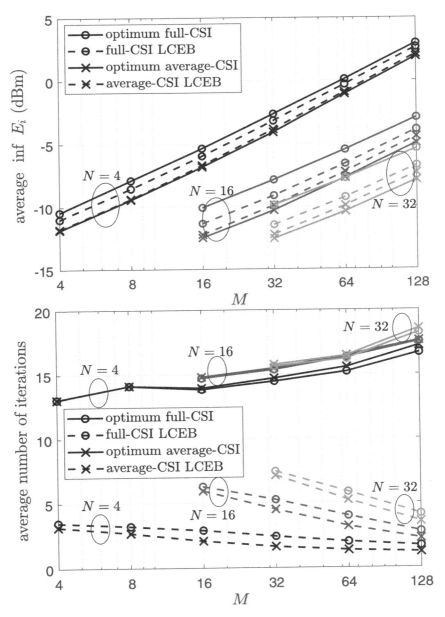

Figure 4.2 (a) Average worst-case RF energy available at the EH devices (top), and (b) average number of iterations (bottom), as a function of the number of PB's antennas for $N \in \{4,16,32\}$.

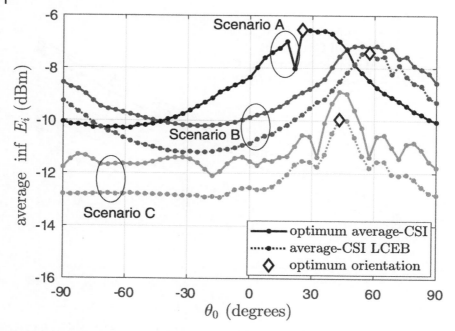

Figure 4.3 Average worst-case RF energy available at the user devices as a function of the PB's rotation angle for three different setups: (i) Scenario A, where $\mathbf{d}$ = [2, 2, 4, 4, 6, 6, 8, 8] m and θ_i = 10i, (ii) Scenario B, where d_i = (1 + i) m and θ_i = 90° − 10i, and (iii) Scenario C, where $\mathbf{d}$ = [3, 3, 5, 5, 7, 7, 10, 10] m and θ = [20, 20, 60, 60, 40, 40, 10, 80]°.

may be considerable as evidenced by Figure 4.3's results. Also, the actual rotation gains are considerable since the gaps between the global minimums and maximums are around 3 dB for the scenarios shown. Meanwhile, the performance gap between the SDP- and LCEB-based designs is not greater than 1 dB for all scenarios (0 dB in case of Scenario A), and their curves follow similar trends.

4.3 CSI-free Multi-antenna Techniques

For the most stringent scenarios, such as massive WET for ultra-low energy devices, CSI-free schemes may be extremely beneficial [78]. The main reasons are that

- CSI-limited EBs, although reduce energy consumption, do not avoid completely the energy expenditure associated to the auxiliary procedures;
- the denser the network is, the more efficient CSI-free schemes may be since the performance of CSI-based systems decays quickly as the number of served devices increases [84].

Based on our previous works [84, 87], we illustrate CSI-free WET with multiple transmit antennas to power efficiently a large set of IoT devices.

4.3.1 System Model and Assumptions

Consider the same scenario introduced in the previous section: a PB equipped with M antennas powers wirelessly a large set S of N single-antenna EH nodes located nearby. Note that in this case, N may be considerably larger than M. Since herein we deal only with CSI-free WET schemes, and for such scenarios the characterization of one harvester is representative of the overall performance, we focus our attention on the case of a generic node S. Therefore, we keep the same notation as in the previous section but avoid the sub-index corresponding to a certain device identity. Without loss of generality, we set the transmit power and the duration of a block to 1. The latter implies that the terms energy and power can be indistinctly used.

We consider quasi-static channels, where the fading process is considered to be constant over the transmission of a block and i.i.d. from block to block, with Rician distribution with factor κ. The PB is assumed to be equipped with a ULA, thus, (4.16)−(4.18) hold here as well. Also, we assume channels with a real covariance matrix $\mathbf{R} \in \mathbb{R}^{M \times M}$ for gaining in analytical tractability, which means that real and imaginary parts of $\hat{\mathbf{h}}$ are i.i.d. and also with covariance $\mathbf{R}$ [228].

Preventive Adjustment of Mean Phases. Since θ, and consequently $\boldsymbol{\phi}$, are different for each harvester, we may not make any preventive phase adjustment based on a specific θ. However, we may still use the geometrical/topological information embedded in (4.18) to improve the statistics of the harvested energy. To explore this, consider that the PB applies a preventive adjustment of the signal phase specified by $\boldsymbol{\psi} \in [0, 2\pi]^{M \times 1}$, and without loss of generality set $\psi_0 = 0$. Then, the equivalent normalized channel vector seen at certain device S becomes $\mathbf{h}' = \Psi \mathbf{h}$, where

$$\Psi = \mathrm{diag}\left(\exp(\mathrm{i}\boldsymbol{\psi}) \right). \tag{4.22}$$

Now, departing from (4.16) and (4.17), we have that

$$\mathbf{h}' = \sqrt{\frac{\kappa}{1 + \kappa}} \Psi \bar{\mathbf{h}} + \sqrt{\frac{1}{1 + \kappa}} \Psi \hat{\mathbf{h}}$$

$$\sim \sqrt{\frac{\kappa}{1 + \kappa}} e^{\mathrm{i}(\boldsymbol{\phi} + \boldsymbol{\psi} + \pi/4)} + \sqrt{\frac{1}{1 + \kappa}} \mathcal{CN}(0, \mathbf{R}), \tag{4.23}$$

which comes from simple algebraic operations and using the fact that $\Psi \hat{\mathbf{h}} \sim \mathcal{CN}(0, \Psi \mathbf{R} \Psi^H) \sim \mathcal{CN}(0, \mathbf{R})$ since Ψ is diagonal with unit absolute values' entries. Now, we rewrite $\mathbf{h}'$ as $\mathbf{h}_x + \mathrm{i}\mathbf{h}_y$, where $\mathbf{h}_x$ and $\mathbf{h}_y$ are independently distributed as

$$\mathbf{h}_{x,y} \sim \sqrt{\frac{1}{2(\kappa + 1)}} \mathcal{N}\left(\sqrt{\kappa}\boldsymbol{\omega}_{x,y}, \mathbf{R}\right), \tag{4.24}$$

where $\boldsymbol{\omega}_{x,y} = \left[1, \cos(\varphi_1 + \psi_1) \mp \sin(\varphi_1 + \psi_1), \cdots, \cos(\varphi_{M-1} + \psi_{M-1}) \mp \sin(\varphi_{M-1} + \psi_{M-1})\right]^T$.

EH Transfer Function. Among the different analytical EH transfer functions surveyed in Table 2.2, we adopt the popular non-linear sigmoidal function [100]

$$g(x) = \frac{\varpi_2(1 - e^{-c_1 x})}{1 + e^{-c_1(x - c_0)}}, \tag{4.25}$$

which is known to describe accurately the non-linearity of EH circuits by properly fitting parameters $c_0 \in \mathbb{R}^+$ and $c_1 \in \mathbb{R}$, while ϖ_2 is the harvested power at saturation.

4.3.2 Positioning-agnostic CSI-free WET

Herein, we overview CSI-free multiple-antenna WET strategies for massive wireless powering. We assume θ uniformly distributed in $[0, 2\pi]$, thus

$$f_\theta(\theta) = \frac{1}{2\pi}, \qquad \theta \in [0, 2\pi], \tag{4.26}$$

which fits scenarios where the θ corresponding to each harvester is unknown, or alternatively, scenarios where there is a very large number of harvesters homogeneously distributed in space such that $f_\theta(\theta) \approx 1/(2\pi)$.

All Antennas Transmitting the Same Signals (AA–SS). Under this scheme, the PB transmits the same signal simultaneously with all antennas and with equal power at each such that the energy harvested by S is given by

$$E^h_{aa-ss} = g(E_{aa-ss}), \tag{4.27}$$

where

$$E_{aa-ss} = \frac{\varrho}{M}\left|\mathbf{1}^T\mathbf{h}'\right|^2. \tag{4.28}$$

In the context of massive WET, this scheme was first proposed in [84] and further investigated in [87].

Theorem 4.2 Conditioned on the mean phase shifts of the powering signals, the distribution of the RF power at the input of the energy harvester under the AA–SS operation is given by

$$E_{\text{aa}-\text{ss}} \sim \frac{\varrho R_\Sigma}{2(\kappa + 1)M} \chi^2 \left(2, \frac{2\kappa\beta(\boldsymbol{\psi}, \theta)}{R_\Sigma}\right),$$

$$(4.29)$$

where $R_\Sigma = \mathbf{1}^T \mathbf{R} \, \mathbf{1}$,

$$\beta(\boldsymbol{\psi}, \theta) = v_1(\boldsymbol{\psi}, \theta)^2 + v_2(\boldsymbol{\psi}, \theta)^2,$$

$$(4.30)$$

$$v_1(\boldsymbol{\psi}, \theta) = \mathbf{1}^T \cos(\boldsymbol{\psi} + \boldsymbol{\phi}) = 1 + \sum_{t=1}^{M-1} \cos\left(\psi_t + \phi_t\right),$$

$$(4.31)$$

$$v_2(\boldsymbol{\psi}, \theta) = \mathbf{1}^T \sin(\boldsymbol{\psi} + \boldsymbol{\phi}) = \sum_{t=1}^{M-1} \sin\left(\psi_t + \phi_t\right),$$

$$(4.32)$$

and $\boldsymbol{\phi}$ is given in (4.18) as a function of θ.

Proof. See Appendix B (Section B.1). $\qquad\qquad\square$

The impact of different phase means on the system performance is strictly determined by $\beta(\boldsymbol{\psi}, \theta)$, and can be better understood by checking the main statistics (mean and variance) of the incident RF power, which can be easily obtained from (4.29) as

$$\mathbb{E}\left[E_{\text{aa}-\text{ss}}|\theta\right] = \frac{\varrho R_\Sigma}{2(\kappa + 1)M} \left(2 + \frac{2\kappa}{R_\Sigma}\beta(\boldsymbol{\psi}, \theta)\right)$$

$$= \frac{\varrho}{M(\kappa + 1)} \left(R_\Sigma + \kappa\beta(\boldsymbol{\psi}, \theta)\right),$$

$$(4.33)$$

$$\mathbb{V}\left[E_{\text{aa}-\text{ss}}|\theta\right] = \frac{\varrho^2 R_\Sigma^2}{4(\kappa + 1)^2 M^2} \left(4 + \frac{8\kappa}{R_\Sigma}\beta(\boldsymbol{\psi}, \theta)\right)$$

$$= \frac{\varrho^2 R_\Sigma}{(\kappa + 1)^2 M^2} \left(R_\Sigma + 2\kappa\beta(\boldsymbol{\psi}, \theta)\right).$$

$$(4.34)$$

Therefore, both mean and variance increase with $\beta(\boldsymbol{\psi}, \theta)$. Meanwhile, it is easy to check that $\beta(\boldsymbol{\psi}, \theta)$ is maximized for $\boldsymbol{\psi} + \boldsymbol{\Phi} = \mathbf{0}$, for which $\beta(\boldsymbol{\psi}, \theta) = M^2$.

Let us assume no preventive adjustment of mean phases is carried out, i.e., $\boldsymbol{\psi} = \mathbf{0}$, to illustrate in Figure 4.4(a) the impact of channel phase means. Specifically, we show $\beta(\mathbf{0}, \theta)$ for different values of M. Note that the number of minima of $\beta(\boldsymbol{\psi}, \theta)$ matches M, thus, as M increases, the chances of operating close to a minimum increase as well, which deteriorates significantly the system performance in terms of average incident RF power.

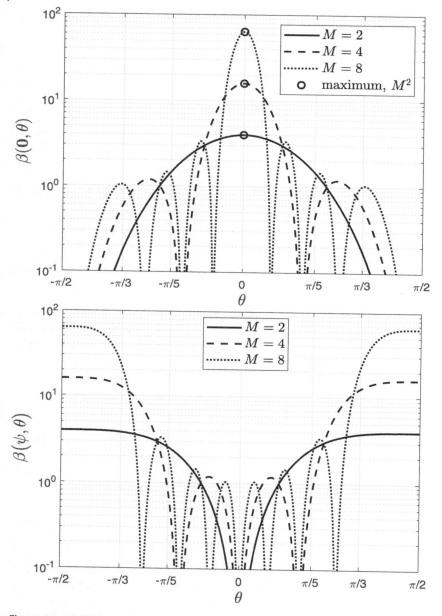

Figure 4.4 (*a*) $\beta(\mathbf{0}, \theta)$ versus θ (top), and (*b*) $\beta(\mathbf{\psi}, \theta)$ versus θ for $\mathbf{\psi}$ according to (4.35) (bottom). We set $M \in \{2,4,8\}$.

Now, we focus on the problem of properly setting the vector $\boldsymbol{\psi}$ for optimizing the system performance.

Theorem 4.3 PB's consecutive antennas must be π phase-shifted, i.e.,

$$\psi_t = \mod (t, 2)\pi, \qquad t \in \{0, 1, \cdots, M - 1\}, \tag{4.35}$$

for optimum average energy performance under the AA–SS scheme.

Proof. We proceed to rewrite (4.30) using (4.31) and (4.32) followed by some algebraic transformations as

$$\beta(\boldsymbol{\psi}, \theta) = \left(1 + \sum_{t=1}^{M-1} \cos (\phi_t + \psi_t)\right)^2 + \left(\sum_{t=1}^{M-1} \sin (\phi_t + \psi_t)\right)^2$$

$$= M + 2 \sum_{t=1}^{M-1} \cos(\phi_t + \psi_t) + 2 \sum_{t=1}^{M-2} \sum_{l=t+1}^{M-1} \cos (\phi_t + \psi_t - \phi_l - \psi_l). \tag{4.36}$$

Then, one needs to solve arg $\max_{\boldsymbol{\psi}} \int_0^{2\pi} \mathbb{E}[E_{aa-ss} | \theta] f_\theta(\theta) d\theta = $ arg $\max_{\boldsymbol{\psi}} \beta(\boldsymbol{\psi})$, where

$$\beta(\boldsymbol{\psi}) = \int_0^{2\pi} \beta(\boldsymbol{\psi}, \theta) f_\theta(\theta) d\theta = \frac{1}{2\pi} \int_0^{2\pi} \beta(\boldsymbol{\psi}, \theta) d\theta, \tag{4.37}$$

which comes from using (4.26). Substituting (4.18) into (4.37) and (4.36), and integrating over θ, we attain

$$\beta(\boldsymbol{\psi}) = M + 2 \sum_{t=1}^{M-1} J_0(t\pi) \cos \psi_t + 2 \sum_{t=1}^{M-2} \sum_{l=t+1}^{M-1} J_0((t - l)\pi) \cos(\psi_t - \psi_l), \tag{4.38}$$

which comes from using the integral representation of $J_0(\cdot)$ [229, Eq.(10.9.1)]. Now, since $|\cos \alpha| \leq 1$, we have that

$$\beta(\boldsymbol{\psi}) \leq M + 2 \sum_{t=1}^{M-1} \left|J_0(t\pi)\right| + 2 \sum_{t=1}^{M-2} \sum_{l=t+1}^{M-1} \left|J_0((t - l)\pi)\right|. \tag{4.39}$$

Using the fact that $J_0(t\pi)$ is positive (negative) if t is even (odd), we can easily observe that the upper bound in (4.39) can be attained by setting ψ_t as in (4.35). $\qquad\square$

In Figure 4.4(b), we show the impact of the above preventive phase shifting on $\beta(\boldsymbol{\psi}, \theta)$ for different angles θ. By comparing Figure 4.4(a) (no preventive phase shifting) and Figure 4.4(b) (preventive phase shifting given in (4.35)), note that 1. the number of minima keeps the same; 2. the best performance occurs now for $\theta = \pm\pi/2$, while the worst situation happens when $\theta = 0$; and 3. there are considerable improvements in terms of area under the curves, which are expected to conduce to considerable improvements when averaging over θ.

Note that although the maximum $\beta(\boldsymbol{\psi})$ as in (4.39) cannot be further simplified, an accurate approximation is given by

$$\beta(\boldsymbol{\psi}) \approx 0.85 \times M^{1.5}, \qquad \text{for } \boldsymbol{\psi} \text{ according to (4.35)}, \qquad (4.40)$$

which comes from standard curve-fitting. Meanwhile, in the case of not introducing phase shifts, i.e., $\boldsymbol{\psi} = \mathbf{0}$, one attains the minimum of β in (4.36) (which minimizes the average incident RF power, although with minimum variance) [87], whose result can be accurately curve-fitted to

$$\beta(\mathbf{0}) \approx 0.64 \times M. \qquad (4.41)$$

We corroborate the accuracy of (4.40) and (4.41) in Figure 4.5.

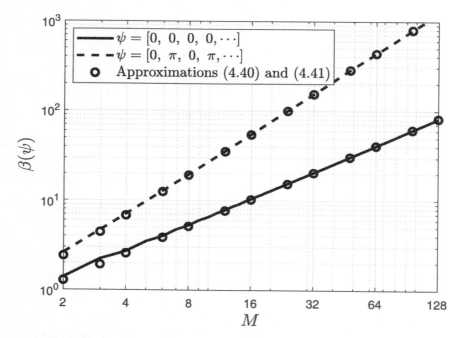

Figure 4.5 $\beta(\boldsymbol{\psi})$ versus M.

Rotary Antenna Beamforming (RAB). The average EH performance under the AA–SS scheme with a certain phase shifting $\boldsymbol{\psi}$ is ultimately determined by the devices' positions through ϱ (see for instance (4.21)) and θ. Without prior CSI and positioning information there is no way to counteract small values of ϱ. Meanwhile, θ may also play a critical role in the EH performance. For instance, devices at (or approximately at) the boresight direction of the PB's antenna array will collect little energy (specially under strong LOS conditions) when the *optimum* preventive phase shifting given in (4.35) is used. Note that by plugging (4.40) into (4.33) one attains just a representative figure of the average incident RF energy counted in all the EH devices with the same path-loss but covering all possible angular directions. Therefore, AA–SS with a certain phase-shifting cannot equally cover all the angular directions, and may become unfair for certain deployed IoT devices. Alternatively, by equipping the PB with a rotary-motor (similar to our discussions in Section 4.2.3, and as in [227]), and letting it continuously rotate, each EH device's angular position will constantly change, thus, allowing each device to experience the average harvested energy claimed for AA–SS in the previous paragraphs. Obviously, it is convenient setting $\boldsymbol{\psi}$ as in (4.35) while performing such endless rotations. We refer to this CSI-free WET scheme as **RAB**.

In practice, a smooth PB's antenna array angular rotation may be difficult to achieve, but it may not even be necessary. Note that the number of minima in the plots of Figure 4.4 matches the number of antennas M, then, at least M angular rotations are needed. In fact, equipping the PB with a stepper motor with M equally-spaced steps already provides a sufficiently smooth performance when using the optimum preventive phase shifting given in (4.35), thus, we adopt such spacing here. As a consequence, a certain device S_i at angular position θ_i will experience an approximate average incident RF energy as that specified in (4.33) but with

$$\hat{\beta}(\boldsymbol{\psi}, \theta_i) = \frac{1}{M} \sum_{j=0}^{M-1} \beta\left(\boldsymbol{\psi}, \theta_i + \frac{j\pi}{M}\right), \tag{4.42}$$

instead of $\beta(\boldsymbol{\psi}, \theta_i)$, which in turn is given in (4.30) or (4.36). As the number of rotation steps increases, $\hat{\beta}(\boldsymbol{\psi}, \theta_i)$ approaches (4.41) or (4.40), when $\boldsymbol{\psi} = \mathbf{0}$ or $\boldsymbol{\psi}$ is set according to (4.35), respectively. Unfortunately, M rotations are still not sufficient when operating without preventive phase shifting as evidenced in Figure 4.6(a). Meanwhile, Figure 4.6(b) shows the favorable performance offered by setting $\boldsymbol{\psi}$ according to (4.35), since this phase shifting maximizes the average gains over a single-antenna transmission, which is specified by $\hat{\beta}(\boldsymbol{\psi}, \theta_i)/M$ for large LOS.

Random Phase Sweeping with Energy Modulation/Waveform (RPS-EMW). This CSI-free WET scheme was proposed in [230]. It may be classified

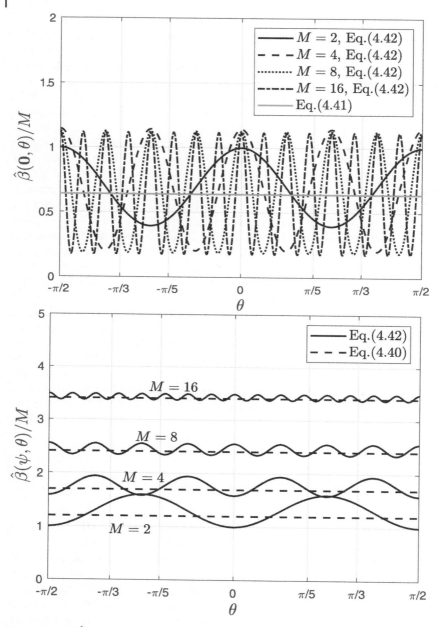

Figure 4.6 (a) $\hat{\beta}(\mathbf{0}, \theta)$ versus θ (top), and (b) $\beta(\boldsymbol{\psi}, \theta)$ versus θ for $\boldsymbol{\psi}$ according to (4.35) (bottom). We set $M \in \{2, 4, 8, 16\}$.

within the AA–SS family in the sense that the same powering signal is still transmitted simultaneously by all the antennas. However, RPS-EMW has some distinctive characteristics: 1. the preventive phase shifts are taken uniformly and independently distributed over $[0, 2\pi]$; and 2. energy modulation/waveform is introduced to provide further enhancements by taking advantage of the non-linearity of the EH circuitry. Specifically, in the case of energy modulation, the energy symbols can be drawn from a distribution with high energy content [231], e.g., circularly symmetric complex Gaussian (CSCG), real Gaussian, flash signaling distribution of [231]; while in case of energy waveform, multi-sine transmissions with uniform power allocation on each antenna may be used. In both cases, there is a significant and comparable performance gain due to the EH non-linearity of the rectenna. Note that different to AA–SS, where the focus was on optimizing the RF energy availability at the EH device side, here the focus lies in optimizing the harvesting process by exploiting the EH circuitry features. Later, when assessing numerically the performance of RPS-EMW, we just refer to the variant with energy modulation and specifically using a real Gaussian distribution.

All Antennas Transmitting Independent Signals (AA–IS). Instead of transmitting the same signal over all antennas, the PB may transmit signals independently generated across the antennas. This is for alleviating the issue of destructive signal combination at S, and it was proposed in the context of massive WET in [87]. We refer to this scheme as AA–IS, for which the harvested energy is given by

$$E^h_{\text{aa−is}} = g(E_{\text{aa−is}}),\tag{4.43}$$

where

$$E_{\text{aa−is}} = \frac{\varrho}{M}||\mathbf{h}'||^2.\tag{4.44}$$

Theorem 4.4 Conditioned on the mean phase shifts of the powering signals, the approximated distribution of the RF power at the input of the energy harvester under the AA–IS is given by

$$E_{\text{aa−is}} \sim \frac{\varrho}{2M^2(\kappa + 1)}\left(R_\Sigma \chi^2\left(2, \frac{2\kappa\beta(\boldsymbol{\psi}, \theta)}{R_\Sigma}\right)\right.$$
$$\left. + \frac{M^2 - R_\Sigma}{M - 1}\chi^2\left(2(M - 1), \frac{2M(M - 1)\kappa\bar{\upsilon}(\boldsymbol{\psi}, \theta)}{M^2 - R_\Sigma}\right)\right),\tag{4.45}$$

where

$$
\tilde{v}(\boldsymbol{\psi}, \theta) = + M - 1 \sum_{j=1}^{M-1} \frac{2}{j(j+1)} \left(\sum_{t=M-j+1}^{M-1} \cos\left(\psi_t + \phi_t\right) \right.
$$

$$
+ \sum_{t=M-j+1}^{M-1} \sum_{l=t+1}^{M-1} \cos\left(\psi_t + \phi_t - \psi_l - \phi_l\right)
$$

$$
- j \cos\left(\psi_{M-j} + \phi_{M-j}\right)
$$

$$
\left. - j \sum_{t=M-j+1}^{M-1} \cos(\psi_{M-j} + \phi_{M-j} - \psi_t - \phi_t) \right). \tag{4.46}
$$

Proof. See Appendix B (Section B.2). ☐

Note that it is not completely clear from (4.45) whether phase shifts are advantageous or not under this scheme.

Let us start by checking the average statistics of E_{aa-is} as follows

$$
\mathbb{E}\left[E_{aa-is}|\theta\right] \approx \frac{\varrho}{2M^2(\kappa+1)} \left(\frac{M^2 - R_\Sigma}{M-1} \left(2(M-1) \right. \right.
$$

$$
\left. + \frac{2M(M-1)\kappa\tilde{v}(\boldsymbol{\psi}, \theta)}{M^2 - R_\Sigma} \right) + R_\Sigma \left(2 + \frac{2\kappa\beta(\boldsymbol{\psi}, \theta)}{R_\Sigma} \right) \right)
$$

$$
= \frac{\varrho}{M^2(\kappa+1)} \left(M^2 - R_\Sigma + M\kappa\tilde{v}(\boldsymbol{\psi}, \theta) + R_\Sigma + \kappa\beta(\boldsymbol{\psi}, \theta) \right)
$$

$$
= \varrho\left(1 + \frac{\kappa\tilde{\beta}(\boldsymbol{\psi}, \theta)}{M^2(\kappa+1)} \right), \tag{4.47}
$$

where $\tilde{\beta}(\boldsymbol{\psi}, \theta) = \beta(\boldsymbol{\psi}, \theta) + M\tilde{v}(\boldsymbol{\psi}, \theta) - M^2$. Notice that the larger $\tilde{\beta}(\boldsymbol{\psi}, \theta)$, the greater $\mathbb{E}\left[E_{aa-is}\right]$. However, $\tilde{\beta}(\boldsymbol{\psi}, \theta) \lll 1$ independently of the value of θ, which can be numerically corroborated. Therefore, channel mean phase shifts do not *strictly* bias the average harvested energy, which is intuitively expected since transmitted signals are independent of each other, and no preventive phase shifting would improve such average statistics [87]. Then, the average incident RF power is approximately ϱ, and this scheme cannot take advantage of the multiple antennas to improve the average statistics of the incident RF power in any way.

Since the average incident RF energy is not affected by any phase shifting, the optimum configuration is the one that minimizes the energy dispersion. In this case, we measure the energy dispersion via variance, which is given by

$$\mathbb{V}[E_{\mathrm{aa-is}}|\theta] \approx \frac{\varrho^2}{2M^4(\kappa+1)^2} \left(\frac{(M^2 - R_\Sigma)^2}{(M-1)^2} \left(2(M-1) \right. \right.$$

$$\left. \left. + \frac{4M(M-1)\kappa\tilde{v}(\boldsymbol{\psi},\theta)}{M^2 - R_\Sigma} \right) + R_\Sigma^2 \left(2 + \frac{4\kappa\beta(\boldsymbol{\psi},\theta)}{R_\Sigma} \right) \right)$$

$$\approx \frac{\varrho^2}{M^3(M-1)(\kappa+1)^2} \left(M^3(1+2\kappa) + R_\Sigma^2 - 2MR_\Sigma(1+\kappa) \right.$$

$$\left. + 2\kappa M(R_\Sigma - M)\beta(\boldsymbol{\psi},\theta) \right), \tag{4.48}$$

where last line comes from taking advantage of $\beta(\boldsymbol{\psi},\theta) + M\tilde{v}(\boldsymbol{\psi},\theta) - M^2 \approx 0 \to \tilde{v}(\boldsymbol{\psi},\theta) \approx M - \beta(\boldsymbol{\psi},\theta)/M$ to write $\mathbb{V}[E_{\mathrm{aa-is}}]$ just as a function of $\beta(\boldsymbol{\psi},\theta)$. Since $\beta(\boldsymbol{\psi},\theta) \geq 0$, it is obvious that just when $M > R_\Sigma$, for instance negative correlation of some antennas, the system performance benefits from having different mean phases.

Then, we have that

$$\arg\min_{\boldsymbol{\psi}} \mathbb{V}[\xi_{\mathrm{aa-is}}^{\mathrm{rf}}] = \begin{cases} \arg\min_{\boldsymbol{\psi}} \beta(\boldsymbol{\psi}), & \text{if } R_\Sigma > M \\ \arg\max_{\boldsymbol{\psi}} \beta(\boldsymbol{\psi}), & \text{if } R_\Sigma < M \end{cases}$$

$$= \begin{cases} \boldsymbol{\psi} \approx \mathbf{0}, & \text{if } R_\Sigma > M \\ \psi_t = \mathrm{mod}(t,2)\pi, & \text{if } R_\Sigma < M \end{cases}, \tag{4.49}$$

where the last line comes from using directly our previous results for AA–SS. In addition, notice that $\mathbb{V}[E_{\mathrm{aa-is}}]$ is not a function of $\beta(\boldsymbol{\psi})$ when $R_\Sigma = M$ (see (4.48)), thus, a preventive phase shifting is not necessary as it does not make any difference. In general, phase shifting is not required since $R_\Sigma \geq M$ holds in most practical systems. On the other hand, observe from (4.48) that the variance decreases with M, thus, although the AA–IS is not more advantageous than single antenna transmissions in terms of the provided average RF energy, it benefits significantly from the multiple antennas to reduce the energy dispersion.

Finally, remember that for AA–SS and AA–IS schemes, the harvested energy comes from mapping the RF energy through the EH transfer function g, (4.25), as shown in (4.27) and (4.43).

Switching Antennas (SA). Instead of transmitting with all antennas at once, the PB may transmit a signal with full power by one antenna at a time such that all antennas are used during a block. This is the SA scheme analyzed in [84] in the context of massive WET.

This is different from AA–SS, RPS-EMW and AA–IS for which the signal power over each antenna is $1/M$ of the total available transmit power and M RF chains were required. In this case just one RF chain is required, hence, reducing circuit power consumption, hardware complexity, and consequently the economic cost.

Assuming equal-time allocation for each antenna, the system is equivalent to that in which each sub-block duration is $1/M$ of the total block duration, and the total harvested energy accounts for the sum of the M sub-blocks. The energy harvested by S is given by

$$E_{sa}^h = \frac{1}{M} \sum_{j=1}^{M} g(E_{sa,j}), \tag{4.50}$$

where

$$E_{sa,j} = \varrho|h_j'|^2 \tag{4.51}$$

is the incident RF power during the jth sub-block.

Notice that for the simple, but commonly adopted in literature, linear EH model, both (4.43) and (4.50) match. However, g is non-linear in practice, and consequently (4.43) may differ significantly from (4.50).

We depart from (4.25) to write the second derivative of g as

$$\frac{d^2}{dx^2}g(x) = \frac{c_1^2 e^{c_1 x}(1 + e^{c_1 c_0})(e^{c_1 c_0} - e^{c_1 x})\varpi_2}{(e^{c_1 c_0} + e^{c_1 x})^3}, \tag{4.52}$$

and note that g is convex (concave) for $x \leq c_0$ ($x \geq c_0$) and $c_1 > 0$, while the opposite occurs for $c_1 < 0$. Let us focus on the case $c_1 > 0$, which is commonly found in practice, then, for certain channel vector realization $\mathbf{h}$ and using Jensen's inequality, we have that

$$g\left(\frac{\varrho}{M} \sum_{j=1}^{M} |h_j|^2\right) \begin{cases} \leq \frac{1}{M} \sum_{j=1}^{M} g(\varrho|h_j|^2), & \text{if } |h_j|^2 \leq \frac{c_0}{\varrho} \; \forall j \\ \geq \frac{1}{M} \sum_{j=1}^{M} g(\varrho|h_j|^2), & \text{if } |h_j|^2 \geq \frac{c_0}{\varrho} \; \forall j \end{cases},$$

$$E_{aa-is}^h \begin{cases} \leq E_{sa}^h, & \text{if } \max|h_j|^2 \leq \frac{c_0}{\varrho} \\ \geq E_{sa}^h, & \text{if } \min|h_j|^2 \geq \frac{c_0}{\varrho} \end{cases}. \tag{4.53}$$

This result implies that devices with EH hardware satisfying $c_1 > 0$ and that are far from the PB, and more likely to operate near their sensitivity level, benefit more from the SA scheme than from AA–IS. However, those closer to the PB and more likely to operate near saturation, benefit more from AA–IS. The analysis is inverted in the case of EH hardware with $c_1 < 0$. All these insights are corroborated in Figure 4.7, where we also appreciate that the statistics of E_{aa-is}^h and E_{sa}^h may be very approximated.

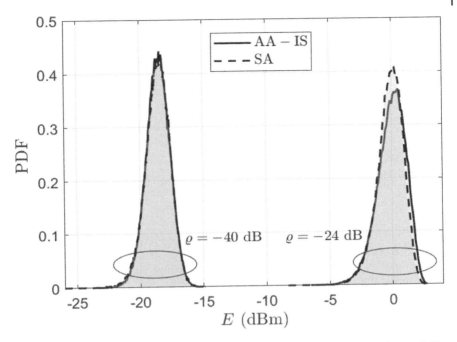

Figure 4.7 PDF of the harvested energy under AA–IS and SA schemes for $\varrho \in \{-40,-24\}$dB. The system is configured as specified in the "Performance evaluation" subsection.

AA–IS and SA schemes provide diversity (full diversity in case of independent channels) even under non-LOS (NLOS) conditions, which does not happen with AA–SS-based schemes. However, it is worth noting that WET channels are usually under some LOS due to the typical WET short distances.

Performance Evaluation. Herein, we show the system performance in terms of average RF/harvested energy in a 20×20 m^2 area served by a multi-antenna PB located in the center, and adopting above positioning-agnostic CSI-free WET schemes. The EH hardware parameters are set as $c_0 = 3.5$, $c_1 = 0.56$ and $\bar{\varpi}_2 = 2$ mW, with sensitivity threshold $\varpi_1 = 0.03$ mW, i.e., $g(x) = 0$, $\forall x \leq \varpi_1$, which agree with the EH circuitry experimental data at 2.45 GHz in [232]. We adopt the path-loss model given in (4.21), and assume the PB is equipped with a four-antenna ULA. Channels are assumed i.i.d., i.e., $\mathbf{R} = \mathbf{I}$, and undergoing Rician fading with $\kappa = 10$. Figure 4.8 evidences that RPS-EMW, SA and AA–IS provide a uniform performance along the area, while AA–SS favors certain spatial directions depending on the adopted $\boldsymbol{\psi}$. Quantitative information on the average harvested energy is shown in Figure 4.9. Observe that while AA–SS without preventive phase shifting allows the EH devices to harvest

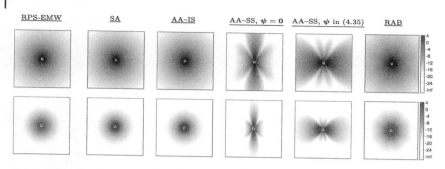

Figure 4.8 Heatmap of the average RF (first row) and harvested (second row) energy in dBm under the discussed CSI-free WET schemes.

energy not less than -14 dBm on average in 10% of the area, such coverage can be increased to 14%, if SA or AA–IS are used, 16% in case of RPS-EMW, or even to 19% under AA–SS with the phase shifting given in (4.35), and 21% when using the rotary AA–SS configuration. It is worth highlighting that in this setup there is no apparent distinction in the performance of SA and AA–IS, while the performance under RPS-EMW deteriorates significantly under higher energy demands. For instance, devices requiring harvesting at least -2 dBm of energy can only accomplish this in 1.6% of the area when the PB uses RPS-EMW or AA–SS without preventive phase shifting, while the coverage area can be extended to 2.5% in case of SA or AA–IS, and up to 3.2% and 3.6% when using the more evolved AA–SS with ψ according to (4.35) and RAB, respectively.

4.3.3 Positioning-aware CSI-free WET

The CSI-free WET schemes discussed in Section 4.3.2 are mostly blind in the sense they exploit little or no information for performance improvements. Meanwhile, positioning information, which may be available in case of static or quasi-static IoT deployments, could be used to improve the WET efficiency. In fact, AA–SS without preventive phase shifting or/and with the preventive phase shifting given in (4.35) can be adapted to partially take advantage of positioning information by properly rotating the PB antenna array. However, a more general and efficient positioning-aware CSI-free scheme may be viable in the case of clustered deployments by adopting specific phase shifts based on the devices positioning information. As emphasized in [78], an efficient design requires an appropriate:

- Clustering algorithm favoring the angular domain (see Appendix C, and a simple pre-processing that may be required for angular-based clustering in Section C.4).

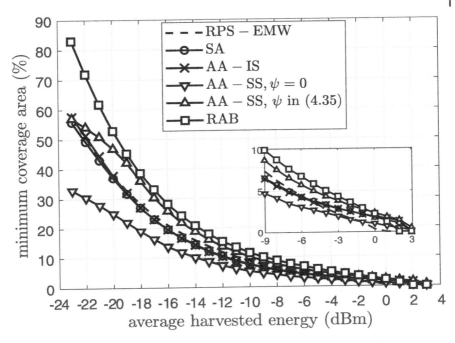

Figure 4.9 Area coverage for different average harvested energy requirements.

- Power control, since clusters with the farthest devices need to be compensated with greater power allocations.
- Per-cluster antenna selection, since the more antennas are used to power certain cluster, the narrower the associated energy beam is. Clusters with greater angular dispersion should be powered using wider beams, i.e., beams generated with a smaller number of antennas.

An example application is illustrated in Figure 4.10 but we do not go into the details of an specific algorithmic implementation.

Note that positioning information is inexact in practice, and the position accuracy must be taken into account. For instance, the research community and industry are targeting a positioning accuracy around 10 cm for IoT deployments in the 6G era. Also, other more practical antenna arrays (instead of the simple ULA) must be intensively analyzed. Finally, although we just emphasized on the need of a location-based clustering, efficient designs may also take advantage (when available) of a device's battery state information and QoS requirements to dynamically reduce the set of devices to be served and optimize the network performance.

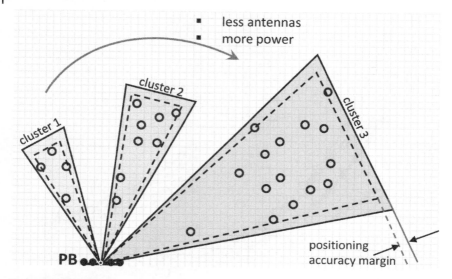

Figure 4.10 Example setup where the IoT deployment can be split into three clusters with 4, 8 and 15 EH devices, and the PB may use a positioning-aware CSI-free EB for efficient WET. In the figure, the inner triangles (with dashed lines) delimit the region where the EH devices are predictably located, while the outer triangles (with solid lines) take into account a positioning accuracy margin. Thus, in the figure we have illustrated the real position of the EH devices. *Source:* Adaptation from [78, Fig. 15].

4.4 On the Massive WET Performance

Herein, we compare the main WET schemes discussed in this chapter in terms of average worst-case EH performance as a function of the number of deployed IoT devices. The simulation setup is the same as that in Section 4.2.3:

- N EH devices randomly and uniformly distributed around the PB at distances between 1 and 10 m, in an annulus region of around 311 m^2 of area;
- a log-distance path-loss model with exponent 2.7, and 16 dB of non-distance dependent loss, such that ϱ_i obeys (4.21) for each S_i;
- $M = 8$, and $\kappa = 10$ for all Rician channels involved.

Additionally, we consider the non-linear sigmoidal EH transfer function given in (4.25), and set the PB transmit power[8] to 3 W. The results are shown in Figure 4.11.

---

8 Different from previous drawn results where the PB transmit power was chosen unitary, herein a higher transmit power is adopted to somewhat counteract the losses in the EH circuitry in the farthest deployed IoT devices.

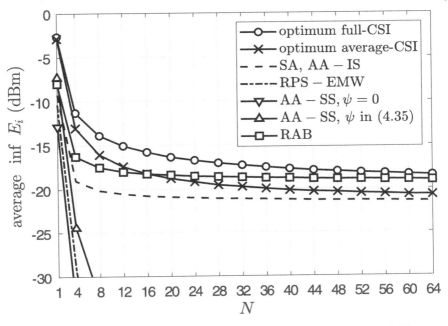

Figure 4.11 Average worst-case harvested energy as a function of the number of EH devices.

Note that the performance under the native AA–SS scheme (with $\boldsymbol{\psi} = 0$ or according to (4.35)) deteriorates quickly as the number of IoT devices increases because of the increasing chances of the latter being located in blind spots of the PB radiation pattern. The other CSI-free WET schemes do not suffer from this since their radiation pattern is angularly uniform (approximately in the case of the RAB scheme). Still, the performance deteriorates quickly also under RPS-EMW because of its more limited radius coverage, while as the number of devices increases, the number of devices far from the PB, which are the ones that mostly determine the worst-case EH performance, increases. Meanwhile, the performance under SA and AA–IS decreases asymptotically to -21.5 dBm, which matches $g(3 \times 10^3/(10^{2.7} \times 10^{1.6})$ mW). As expected, the optimum full-CSI scheme outperforms all its counterparts but the performance gap with respect to the RAB, AA–IS and SA decreases as more devices require to be powered. This is without considering the power consumed in the CSI acquisition, which also may increase with N and would tilt the scale in favor of the CSI-free/-limited schemes. Interesting, the AA–SS scheme optimized via rotation capabilities can outperform the optimum average-CSI scheme already for $N > 16$ and quickly approaches the performance of the optimum full-CSI scheme. Finally, note that the performance under the optimum average-CSI approach deteriorates quickly as the number of deployed IoT devices N grows above the available number of transmit antennas M.

4.5 Final Considerations

In this chapter, we looked into EH from dedicated WET. We analyzed a single-PB setup, and discussed diverse transmit schemes for optimizing the WET process. Specifically, we addressed EB techniques with full, limited, or even without, CSI that the PB may utilize for powering a massive number of EH single-antenna IoT devices. Note that novel CSI-limited/-free EBs are key to supporting massive WET since the gains from traditional full and perfect CSI-based EB may not compensate the energy consumed during the CSI acquisition procedures.

We illustrated an attractive CSI-limited solution that makes use of only the average channel statistics, and as such, it does not require frequent CSI updates, it can be learned with limited CSI-acquisition overhead and energy expenditure, and it is less prone to estimation errors. The performance under such average-CSI EB was shown to approach the full-CSI counterpart as the LOS component in the WET channels becomes stronger. Moreover, the devices' worst-case harvesting performance was found to follow a close-to-linear increasing trend as the number of PB's antennas increases. Unfortunately, the optimum full-CSI and average-CSI EBs come from solving SDP problems, which become computationally costly in the mMIMO regime. To sort this out, we presented a near-optimum low-complexity EB design for both full and average CSI situations.

The average-CSI EB may offer significant energy savings and may be attractive in many IoT use cases. However, energy expenditure is not completely avoided and also the system performance may deteriorate significantly when the number of EH devices becomes much larger than the number of PB's antennas. To address such challenges we discussed several state-of-the-art CSI-free WET schemes and highlighted their pros and cons. Since such WET schemes are mostly blind in the sense they exploit little or no information for performance improvements, we defended the idea of exploiting positioning information, which may be available in the case of static or quasi-static IoT deployments, and even devices' battery state information and/or associated QoS requirements, to improve the WET efficiency. Finally, throughout this chapter, we highlighted and evidenced the prominent benefits of incorporating rotary capabilities into the PB.

5

Multi-PB Massive WET

5.1 On the PBs Deployment

As discussed in Chapter 2, DAS or multi-PB systems are particularly attractive for banning RF energy delivery blind spots in relatively-wide EH areas. Let us consider a set $\mathcal{T} = \{T_1, T_2, \cdots, T_{N'}\}$ of N' PBs that are available to be deployed in a certain area $\mathcal{A} \subset \mathbb{R}^2$ to energize a set $\mathcal{S} = \{S_1, S_2, \cdots, S_N\}$, of N EH devices. We assume two cases. In the first, we consider all PBs and EH devices equipped with a single antenna for simplicity and ease of tractability. However, later in this chapter, we assume multi-antenna PBs and discuss EB-based strategies and related gains. The deployment of PBs and devices is given by the set of position coordinates $\{c_j^t = (x_j^t, y_j^t)\}$, and $\{c_i^s = (x_i^s, y_i^s)\}$, respectively.

The PBs deployment problem lies specifically in *finding the appropriate PBs deployment coordinates $\{c_j^t\}$ given QoS requirements and system constraints, with or without the EH devices' positioning information.* We focus on maximum fairness problems where all EH devices share similar energy demands (similar to **P2** in Section 4.2) and considering a normalized total transmit power budget constraint. Next, we discuss the particularities of such problems and related solving strategies when the positioning information of EH devices is (Section 5.1.1) or is not (Section 5.1.2) available.

5.1.1 Positioning-aware Deployments

Efficient deployment strategies that aim to maximize the minimal energy harvested E_i require exploiting accurate path loss models for the area of interest. Let us say the path loss between coordinates c_j^t and c_i^s is given by $\varrho_{i,j}$, and it is accurately known once c_j^t has been fixed. Then, the optimization problem can be formulated as **P5** in (5.1), where p_j is the transmit power of T_j.

Wireless RF Energy Transfer in the Massive IoT Era: Towards Sustainable Zero-energy Networks.
First Edition. Onel Alcaraz López and Hirley Alves.
© 2022 by The Institute of Electrical and Electronics Engineers, Inc. Published 2022 by John Wiley & Sons, Inc. Companion website: www.wiley.com/go/Alves/WirelessEnergyTransfer

$$
\textbf{P5: } \max_{\{c_j^t, p_j\}} \quad \inf_{i=1,\cdots,N} E_i = \inf_{i=1,\cdots,N} \sum_{j=1}^{N'} p_j \varrho_{i,j} \tag{5.1a}
$$

$$
\text{s.t.} \quad \sum_{j=1}^{N'} p_j \leq 1. \tag{5.1b}
$$

This problem is highly non-linear because of the tangled dependence between $\varrho_{i,j}$ and $\{c_j^t, c_i^s\}, \forall i, j$. Herein, we adopt the log-distance path-loss model

$$
\varrho_{i,j} = 10^{-1.6} \times d_{i,j}^{-2.7}, \tag{5.2}
$$

exactly as in (4.21), and note that $d_{i,j} = ||c_j^t - c_i^s|| = \sqrt{(x_j^t - x_i^s)^2 + (y_j^t - y_i^s)^2}$. To the best of our knowledge, this problem cannot be rewritten in a convex form [220], thus, a global optimum solution is not guaranteed by any solving method. However, modern meta-heuristics methods such as genetic algorithms (GAs) and particle swarm optimization (PSO) usually provide near-optimum results at the cost of computational complexity, and given the associated hyperparameters are carefully selected. This may be admissible in static or quasi-static setups where the optimization does not need to be run often, and as long as the PB's hardware/software features allow it, or a central controller assists the process.

On the other hand, under strict computing limitations, and/or in more dynamic scenarios where PBs can re-adjust/update their positions, e.g., in the case of moving/flying PBs (see Section 2.4.4), according to the EH device's dynamics, lighter optimization approaches are desirable. In this regard, clustering-based solutions seem a natural choice for dealing with the PBs deployment problem since the influence that a PB has on the surrounding perimeter area decreases quickly, following a power-law, as the area radius increases.

Appendix C surveys the main clustering approaches with special emphasis on K-Means methods. The commonly used K-Means, with Euclidean distance measure and centroids computed as the mean of the points of the associated clusters, seems appealing since the path loss, which dominates the system performance, depends strictly on the Euclidean distance between PBs and EH devices. Even so, next we propose some modifications that can be incorporated to the traditional K-Means method to improve the system performance for the discussed scenario.

Equal power allocation per PB is assumed in a first instance, while later we specifically address the power control problem. This means that **P5** is decomposed into two sub-problems: 1. PBs positioning, and 2. power control.

K-Means with Chebyshev Centroids (K-Chebyshev). As discussed in Appendix C (Section C.1.1), at each iteration of the traditional K-Means method, each data point (EH device) is assigned to its closest cluster head/center (PB), while the clusters centers are re-computed right after. The procedure ends once the cluster centers remain unchanged. In the case of PBs deployment for homogenizing the RF energy transfer to the entire set of EH devices, updating the clusters' centers via the mean of their associated points as in the traditional K-Means methods does not seem completely appropriate. To maximize the minimum energy transfer guarantees within each cluster, the best position for each cluster head is the Chebyshev center[1] of the clusters' points. Formally, let $S_j \subset S$ be the set of points (EH devices) clustered around the cluster head (PB) $T_j \in \mathcal{T}$ at a certain iteration of the K-Means clustering, then, c_j^t can be updated as

$$c_j^t = \arg \min_{x \in \mathbb{R}^2} \max_{S_i \in S_j} \{||c_i^s - x||^2\}. \tag{5.3}$$

The optimization problem in (5.3) can be easily written in convex form as

P6: $\min_{r \in \mathbb{R}, c_j^t \in \mathbb{R}^2} \quad r$ (5.4a)

s. t. $\quad ||c_i^s - c_j^t||^2 - r \leq 0, \qquad \forall S_i \in S_j,$ (5.4b)

where $\sqrt{r}$ is the radius of the minimal-radius circumference enclosing the positions of all the devices belonging to S_j. **P6** can be solved by standard convex optimization tools, such as CVX [234], or dedicated optimization algorithms as that proposed in [233].

Figure 5.1 exemplifies an IoT deployment scenario with $N = 64$ EH devices clustered around $N' = 8$ PBs. PB positions are set using K-Means and K-Chebyshev approaches, which are shown to provide different deployment results.

---

1 There are several definitions of the Chebyshev center of a bounded set $\mathcal{B}$, e.g., 1. the center of the minimal-radius ball enclosing the entire set $\mathcal{B}$ [233], or alternatively (and non-equivalently) 2. the center of the largest inscribed ball of $\mathcal{B}$ [220]. Herein, we refer to the former.

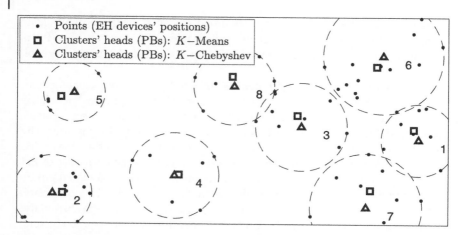

Figure 5.1 Examples of clustering for a scenario with $N = 64$ EH devices and $N' = 8$ PBs. Each cluster is enclosed within its Chebyshev circumference, and its (randomly selected) corresponding index is illustrated.

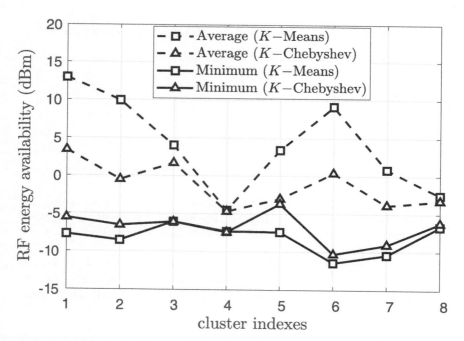

Figure 5.2 Performance of K-Means versus K-Chebyshev clustering in terms of RF energy availability for the scenario illustrated in Figure 5.1 with dimensions 30 m × 15 m. Per-PB transmit power is set to $1/N' = 1/8$W.

K-Means refers to the traditional approach with Euclidean distance and centers of clusters computed as the mean of the associated points, while K-Chebyshev differs from the previous only in the computation method of the cluster centers, for which it uses the Chebyshev center of the cluster's points. Meanwhile, Figure 5.2 shows the system performance in terms of RF energy availability under each of these clustering approaches. Observe that K-Means favors the average energy availability, while K-Chebyshev guarantees a fairer distribution of the energy resources.

Power Allocation. Once the PB positions have been determined, the corresponding power allocation can be easily and optimally found by solving the LP **P7** in (5.5).

$$\textbf{P7: } \min_{\xi,\{p_j\}} \quad -\xi \tag{5.5a}$$

$$\text{s. t.} \quad E_i = \sum_{j=1}^{N'} p_j \varrho_{i,j} \geq \xi, \qquad i = 1, \cdots, N \tag{5.5b}$$

$$\sum_{j=1}^{N'} p_j \leq 1. \tag{5.5c}$$

Note that **P7** comes from simple transformations of **P5** for known $\{c_j^t\}$, thus, known $\{\varrho_{i,j}\}$. This problem is constituted by $N' + 1$ variables and $N + 1$ constraints, where $N \gg N'$ holds usually in practice. If IPMs are used, solving **P7** will take at most $\mathcal{O}(\sqrt{N+1}\log(1/\varepsilon))$ iterations, each one with at most $\mathcal{O}((N+1)^3)$ arithmetic operations [221]. Note that ε represents the solution accuracy attained when the algorithm ends.

As clusters become more separated, the solution of **P7** will tend to be a power allocation proportional to the path loss of the farthest EH device in the cluster, i.e.,

$$p_j^{\text{opt}} \approx \frac{\max_i\left(\varrho_{i,j}^{-1}\right)}{\sum_{j=1}^{N'} \max_i\left(\varrho_{i,j}^{-1}\right)}. \tag{5.6}$$

This phenomenon is illustrated in Figure 5.3.

Overall Performance. Next we discuss the overall performance of the aforementioned approaches for PBs deployment optimization. MATLAB optimization solvers `particleswarm`, `kmeans`, `linprog` are used to draw the results

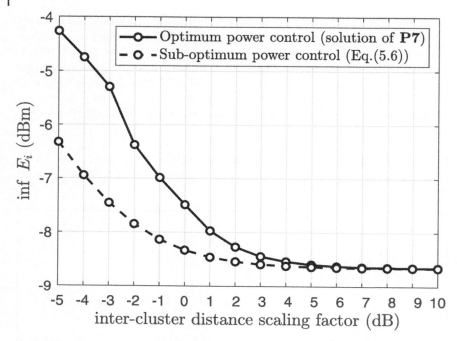

Figure 5.3 Minimum RF energy availability as a function of the inter-cluster distance scaling factor. Herein, we adopt the example scenario in Figure 5.1 (using K-Chebyshev) with 30 m × 15 m as reference, i.e., 0 dB inter-cluster scaling factor.

corresponding to PSO, K-Means and the solution of **P6**, respectively. Also, the cluster centers computation subroutine of kmeans needs to be modified to obey the solution of **P6** and realize the K-Chebyshev clustering. Numerical results as a function of the number of deployed PB and EH devices are illustrated in Figures 5.4, and are drawn assuming the EH devices are randomly deployed in a 30 m × 15 m rectangular area. Observe the expected increasing/decreasing performance as the number of PBs/EH devices increases. Also, note that the PSO solver outperforms its clustering competitors in most cases, especially as the number of devices increases since the clustering patterns tend to vanish. In the scenario with N = 64 EH devices, PSO was just outperformed by K-Means and K-Chebyshev for $N' \geq 10$ and $N' \geq 14$, respectively. However, by increasing the number of PSO particles, and consequently conducting a heavier optimization, PSO is expected to excel also here and in every scenario.

5.1.2 Positioning-agnostic Deployments

When PBs are to be deployed to remain static, while the EH IoT deployments are dynamic (as in case of moving EH devices) or positioning information is

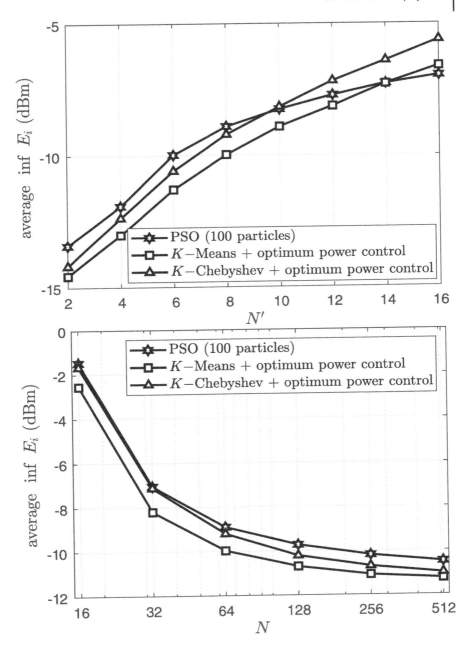

Figure 5.4 Average worst-case RF energy available at a set of *N* EH devices powered by *N'* PBs. Performance is shown as a function of (a) the number of deployed PBs *N'* for *N*=64 (top), and (b) the number of deployed EH devices for *N'* = 8 (bottom). The EH devices are randomly deployed in a 30 m × 15 m rectangular area.

not available, the deployment needs to be carried out to optimize the area RF energy availability performance. This is the case, for instance, of the work carried out in [88], where the authors aim to maximize the minimum average RF energy that is foreseen to be available in the service area.

One may think that it is just enough to distribute PBs uniformly over the service area. A problem that may be analogous to the *circle packing* mathematical formulation [235]. However, the algorithmic and computational complexity of this approach is challenging, especially if service areas are irregular, which is common in practice (including the presence of obstacles where EH devices are surely not deployed). There is no known circle packing algorithm that works for any area shape. Moreover, note that more centered deployed PBs would impact more significantly distant regions than those deployed closer to the service area contour. Therefore, although a uniform deployment of PBs over the service area may lead to near-optimum results in certain scenarios, it is not strictly optimum, especially in terms of fairness as considered here. A circle packing algorithm adapted to pack circles with position-dependent dimensions (as required for strictly optimum performance) is desired, but unfortunately extremely cumbersome to design. Alternatively, when PBs are subject to global power budget constraints instead of per-PB power constraints, a proper global power allocation may alleviate the issue.

Optimization Framework. Fortunately, in general, the optimization approaches discussed in the previous subsection can be re-utilized here by using the artifice of virtually deploying a large number of points (virtual EH devices) in the service area where the received RF energy is then estimated. In addition, herein we propose another optimization approach given in Algorithm 2, which although it can be used given a prior devices' positioning information (as in Section 5.1.1), its potential is fully realized in the asymptotic regime where the number of devices is very large, or equivalently, the positions of devices are unknown.

The proposed greedy iterative movement-based PBs deployment (GIM) optimization algorithm works as follows. First, PB positions, and the set $\mathcal{T}^\circ$ of PBs that are allowed to update their positions, are initialized (lines 2 and 3). Then, the position of each PB in $\mathcal{T}^\circ$ is iteratively updated by 1. estimating the worst performing virtual node in the area (lines 5, 6), 2. determining the corresponding closest PB that is allowed to update its position (line 7), and 3. moving such PB a distance μ (specified as input of the algorithm) toward the virtual node (lines 8 and 9). Once a PB updates its position, it is removed from the set of those that are still eligible (line 10), and only when all PBs have sequentially updated their positions, $\mathcal{T}^\circ$ is reset to include again all the PBs (lines 11-13). Note that the value of μ must be area-dependent and requires careful selection. The larger μ is, the faster the algorithm convergences, but the resulting PBs deployment may be extremely sub-optimal.

Algorithm 2 Greedy iterative movement-based PBs deployment (GIM)

1: **Input:** $\{c_i^s\}$ and μ
2: Initialize $\{c_j^t\}$, e.g., set $c_j^t = \mathbb{E}_i[c_i^s]$, $\forall j$
3: $\mathcal{T}^\circ \leftarrow \mathcal{T}$
4: **repeat**
5: $\quad E_i = \sum_{j=1}^{N'} \varrho_{i,j}$, $\forall i$
6: $\quad i^* = \arg\min_i E_i$
7: $\quad j^* = \arg\min_{j | T_j \in \mathcal{T}^\circ} d_{i^*,j}$
8: $\quad \vartheta = \mathrm{atan2}(c_{i^*}^s - c_{j^*}^t)$
9: $\quad c_{j^*}^t \leftarrow c_{j^*}^t + (\mu\cos\vartheta, \mu\sin\vartheta)$
10: $\quad \mathcal{T}^\circ \leftarrow \mathcal{T}^\circ \setminus T_{j^*}$
11: $\quad$ **if** $|\mathcal{T}^\circ| = 0$ **then**
12: $\quad\quad \mathcal{T}^\circ \leftarrow \mathcal{T}$
13: $\quad$ **end if**
14: **until** convergence
15: **Output:** $\{c_j^t\}$

Numerical Results. Figure 5.5 shows the energy heatmap coverage when $N' = 8$ PBs are deployed according to the aforementioned optimization approaches in a 30 m × 15 m rectangular area. Note that by incorporating power control to the traditional circle packing approach (Figure 5.5(a)), performance gains in the order of ~ 1 dB were attained by allocating less power to the most centered PBs (Figure 5.5(b)), while the gains increased to ~ 2 and ~ 3 dB when using PSO (Figure 5.5(c)) and the proposed GIM framework (Figure 5.5(d)), respectively. In case of the latter, and since the per-PB transmit power is kept fixed, the gains come from deploying some PBs closer to the area edges. Observe also the virtual movement followed by the PBs during the iterative execution of the GIM approach. Moreover, a quantitative statistical information on the energy availability in the entire service area is illustrated in Figure 5.6. Note that our proposed scheme guarantees delivering -10 dBm of RF power to 96% of the service area, while the coverage decreases to 95%, 94% and 83% when using circle packing without/with power control and the PSO approach, respectively. For greater energy requirements, the traditional circle packing algorithm outperforms its competitors since the optimization focus of the latter is on the worst-performing area position, while no performance constraints/utility incentives are imposed/assigned to the remaining positions.

The number of points virtually deployed in the area before running the previously discussed optimization approaches impacts heavily the performance results. The greater the number of measurement points (or virtual EH

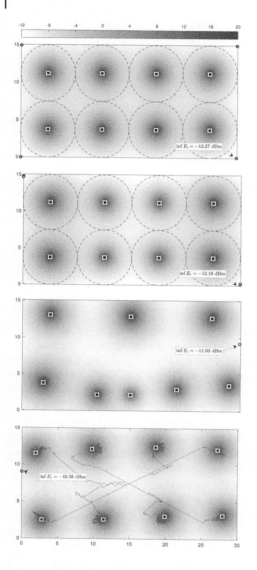

Figure 5.5 Heatmap of the RF energy availability in a 30 m × 15 m area under different strategies for the deployment of $N' = 8$ PBs (illustrated with square white-edge markers): (a) circle packing with equal power per PB (top), (b) circle packing with optimum power control (middle-top), (c) PSO with 100 particles (middle-bottom), and (d) GIM with $\mu = 0.2$ (bottom). The packed circles are shown in case of (a) and (b), while in (d) we illustrate the iterative virtual movement followed by the PBs. In the latter, we adopted as stopping criterion a relative error of 10^{-6} in $\min E_i$ from one iteration to another. We deployed a uniform grid of 60 × 30 virtual points for implementing the power control in (b), and the strategies in (c) and (d).

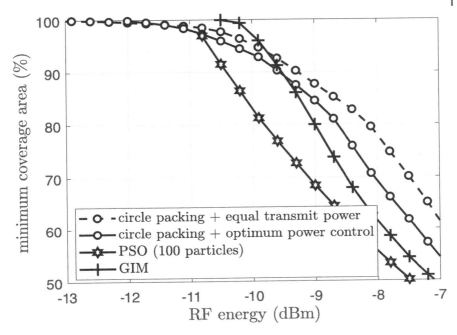

Figure 5.6 Complementary CDF (CCDF) of the area RF energy coverage for the scenario and optimization strategies illustrated in Figure 5.5.

devices), the better the performance results as illustrated in Figure 5.7, although convergence becomes slower since the optimization complexity increases. This does not apply to the traditional circle packing approach (circle packing + equal transmit power) which does not require deploying virtual points. On the other hand, observe the performance gains from incorporating power control to the circle packing approach, and from using more PSO particles as the number of virtual points increases. Interestingly, our proposed GIM framework outperforms all its competitors when deploying a sufficiently large number of measurement points.

Finally, note that if some statistical information on the EH device's positioning is available, e.g., the devices' deployment PDF, it can be used to generate the virtual points, and even schedule moving trajectories, with specific time-varying pauses in some locations, in cases of moving/flying PBs. In this case, AI-based methods can be handy here as well.

5.2 Multi-antenna Energy Beamforming

In the previous section, single-antenna PBs were considered. However, in practice, PBs are often multi-antenna devices with moderate-to-high signal processing capabilities. By exploiting such spatial degrees of freedom,

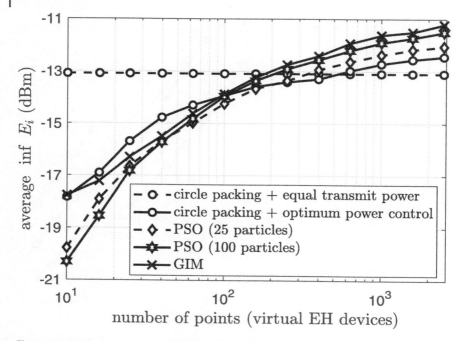

Figure 5.7 Average worst-case RF energy available in the area[2] as a function of the number of virtual points randomly deployed to execute the optimization approaches.

significant performance gains can be obtained via EB or even intelligent CSI-free schemes as discussed in the previous chapter. Consequently, herein we assume each PB is equipped with M antennas and exploits either global (centralized) or local (distributed) CSI for EB. Note that the former is associated to a coordinated multi-point WET networking architecture, while the latter is related to a locally-coordinated design. The position of the PB is assumed to be given beforehand, for instance as a result of placement optimization as discussed in the previous section.

5.2.1 Centralized Energy Beamforming

In a *centralized* scheme, each PB must acquire the CSI of the channels towards all EH devices in the area. Let us denote by $\mathbf{h}_{i,j} \in \mathbb{C}^{M \times 1}$ the small-scale fading channel coefficients between the M antennas of T_j and S_i, thus, $\mathbf{h}'_{i,j} = \sqrt{\varrho_{i,j}} \mathbf{h}_{i,j}$ is the complete CSI vector. This information is forwarded to a *central unit*, which manages the global CSI of the system.

The beamforming coefficients for each PB are computed by the *central unit* as follows. First, set $\mathbf{h}'_i$ as the vertical concatenation of channel vectors

---

2 A total of another 10^4 virtual points were used to measure the RF energy availability in the area.

$\mathbf{h}'_{i,1}, \mathbf{h}'_{i,2}, \cdots, \mathbf{h}'_{i,N'}$. Then, the EB optimization can be carried out in the same way as for a single PB setup. Specifically, the optimization framework in Section 4.2 can be re-utilized by just substituting $\varrho_i \leftarrow 1$, $\mathbf{h}_i \leftarrow \mathbf{h}'_i$, $\forall i$, and noticing that the number of antennas is now equal to $M \leftarrow M \times N'$.

Under ideal conditions, i.e.,

1. perfect, instantaneous, and cost-free CSI acquisition,
2. high-capacity control channels between PBs and the central unit,

centralized EB provides the optimum performance. The latter condition may not be an issue for novel DAS deployments such as the radio stripes system (see Section 2.4.3). However, in general, above ideal conditions are not always nearly met, especially when serving massive deployments of low-power EH IoT devices, thus, paving the way to distributed implementations.

5.2.2 Distributed Energy Beamforming

In a *distributed* scheme, each PB requires acquiring only the CSI of channels towards its surrounding/associated EH devices. There is no central unit, thus, each PB decides its own beamforming strategy based on the available local CSI. However, note that when the entire system is subject to a total power budget, a certain coordination among the PBs may be required to agree on the power resources to be used by each PB. In such scenario, the per-PB transmit power can be allocated by solving **P7**, or simply evaluating (5.6), for maximum fairness WET.

Let us denote by $S_j \subset S$, the set of EH devices that are being intentionally powered by T_j. Note that $\sum_{j=1}^{N'} |S_j| \geq N$, where the equality is strictly attained when every EH device is being intentionally served by only one PB. Herein, we just focus on the latter case. Now, since each PB is responsible for powering a smaller set of EH devices, the CSI acquisition problem is seriously alleviated. Then, the EB optimization can be carried out independently for each PB and its associated EH devices in the same way as for a single PB setup discussed in Section 4.2.

5.2.3 Available RF Energy

Let $\mathbf{w}_{k,j}$ be the precoding vector applied by T_j to its kth transmitted energy signal, and set $\mathbf{W}_j = \sum_{\forall k} \mathbf{w}_{k,j}\mathbf{w}_{k,j}^H$. Then, by analogy to (4.2) and (4.4), the RF energy (normalized by unit time) available at each S_i is given by

$$E_i = \sum_{j=1}^{N'} \mathrm{Tr}\left(\mathbf{W}_j\mathbf{h}'_{i,j}\mathbf{h}'^H_{i,j}\right) = \sum_{j=1}^{N'} \varrho_{i,j} \mathrm{Tr}\left(\mathbf{W}_j\mathbf{h}_{i,j}\mathbf{h}^H_{i,j}\right). \tag{5.7}$$

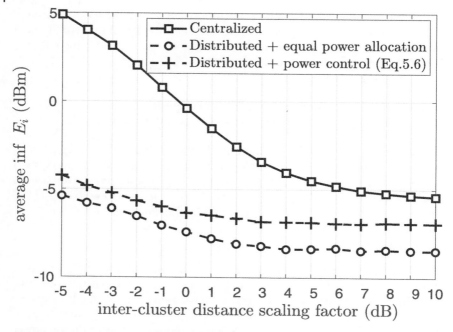

Figure 5.8 Average worst-case RF energy available at a set of N = 64 EH devices powered by N' = 8 PBs, each one equipped with M = 4 antennas, as a function of the inter-cluster distance scaling factor (similar to our exposition in Figure 5.3) and under the centralized and distributed EB designs.

A comparison between the performance under centralized and distributed EB is illustrated in Figures 5.8 and 5.9, where we assumed that each PB is equipped with a ULA and channels undergo Rician fading with LOS factor of 10 dB. Specifically, Figure 5.8 shows an analysis similar to that in Figure 5.3 and evidences that as the network scales larger the distributed EB becomes increasingly appealing, especially when using a proper per-PB power allocation strategy. However, the gains from an ideal centralized EB implementation can remain substantial for many different scenarios, which is also illustrated in Figure 5.9. Observe that while at least 1 mW of average RF power may be available at the EH circuit input of the IoT devices when using centralized EB and M = 8 antennas per PB, the availability decays to 0.26 mW and 0.12 mW in cases of using distributed EB with and without power control, respectively. However, the performance gaps between the different approaches remain approximately constant as the number of transmit antennas increases.

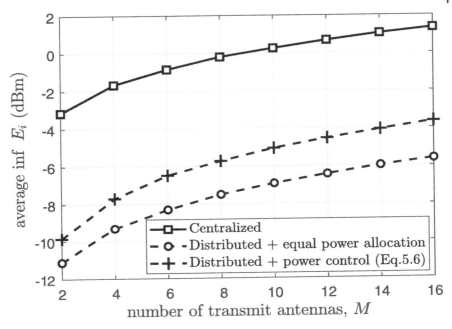

Figure 5.9 Average worst-case RF energy available at a set of $N = 64$ EH devices powered by $N' = 8$ PBs as a function of the number of transmit antennas M, and under the centralized and distributed EB designs. The EH devices are randomly deployed in a 30 m × 15 m rectangular area.

5.3 Distributed CSI-free WET

In previous sections we analyzed separately the single-antenna PBs deployment (with and without devices' positioning information), and EB for multi-antenna PBs problems. In general, these two problems cannot be coupled since EB commonly requires some kind of CSI, which is naturally time-varying. However, when the multi-antenna PBs are deployed to operate without any CSI, an efficient positioning strategy may still take advantage of some additional information, which depends on the adopted CSI-free WET scheme. Next, we discuss how the PBs deployment can be implemented when using the CSI-free schemes discussed in Section 4.3.

5.3.1 SA, AA–IS and RPS–EMW

SA, AA–IS and RPS-EMW have something in common: they all provide an omnidirectional radiation pattern with diversity, but no average gains, concerning single antenna WET. Therefore, the deployment when any of these strategies is used can be done similarly as illustrated previously in Sections 5.1.1 and 5.1.2 for single-antenna PBs.

5.3.2 AA–SS

Under the influence of some LOS, AA–SS favors certain spatial directions depending on the adopted antenna phase shift vector $\psi \in [0, 2\pi]^{M \times 1}$. Then, for a given phase shift configuration, the PBs can be conveniently deployed such that the corresponding radiation patterns efficiently cover the set of EH devices and/or locations of interest.

Similar to our discussions in Section 4.3, herein we focus on the scenario where the PBs are equipped with ULAs. In such cases, it was shown that by using $\psi = 0$ and ψ according to (4.35), a PB's radiation pattern focuses around $\theta = 0$ and $\theta = \pi/2$ radians of the boresight of the antenna array, respectively (see Figure 4.4). This roughly means that in an efficient deployment, most of the EH devices, and/or the more distant ones, should respectively remain around such angular positions with respect to the PB as illustrated in Figure 5.10.

The illustration in Figure 5.10 suggests that it might be appropriate identifying device clusters and deploying a PB between two nearby clusters while properly setting its angular position according to the phase shift configuration. From here on, we set ψ according to (4.35) since this configuration provides the wider radiation pattern. Then, if N' PBs are to be deployed, one would need to identify $2N'$ device clusters.

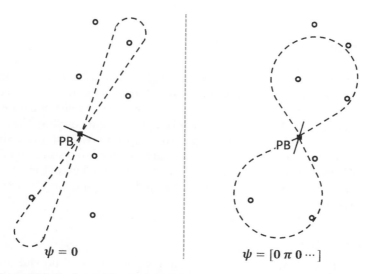

Figure 5.10 Optimum PB orientation for powering a given set of EH devices when using (a) $\psi = 0$ (left) and (b) ψ according to (4.35) (right).

Clustering. In the following, we consider two main alternatives to implement such clustering.

- **Cluster pairing:**
  Cluster pairing consists of identify first $2N'$ device clusters, which can be then paired in

  $$
  \begin{aligned}
  F &= (2N' - 1)(2N' - 3)\cdots 3 \times 1 \\
  &= \frac{(2N' - 1)(2N' - 2)(2N' - 3)\cdots 3 \times 2 \times 1}{(2N' - 2)(2N' - 4)\cdots 2} \\
  &= \frac{(2N' - 1)!}{2^{N'-1}(N' - 1)!}
  \end{aligned}
  \tag{5.8}
  $$

  different ways. For illustration purposes, herein we adopt two different pairing approaches:

  - min max pairing: cluster pairs are selected to minimize the maximum distance between paired cluster centers; and
  - min mean pairing: cluster pairs are selected to minimize the mean distance between paired cluster centers.

  Unfortunately, the cluster pairing approach is computationally costly and often prohibitive as it requires executing a large number of combination/permutation operations as N' increases. In fact, such an operation already becomes infeasible for $N' \geq 5$, thus, we evaluate these approaches only for $N' \leq 4$.

- **Cluster splitting:**
  First, N' clusters are identified. Then, the devices in each of them are again divided into two different clusters. This approach is obviously simpler than those above and computationally affordable. The resultant clustering is the same as in case of single antenna PBs deployment (or PBs with omnidirectional pattern in general), but cluster centers, and ultimate PBs positions, may vary.

Figure 5.11 illustrates how the PBs deployment looks when using the above approaches and K-Chebyshev clustering. PB locations are set in the middle of the associated cluster centers.

PB Orientation. Finally, the orientation of each PB is set such that its boresight direction is aligned (for $\psi = 0$) or perpendicular (for ψ according to (4.35)) to the line connecting the centers of the paired clusters as shown in Figure 5.12.

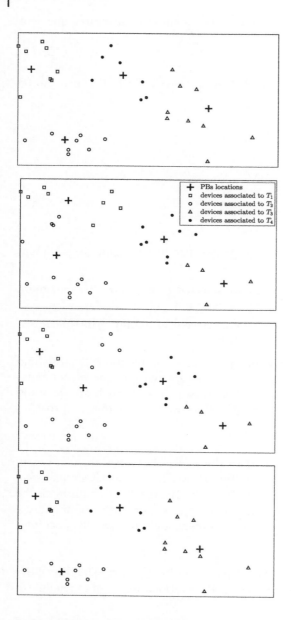

Figure 5.11 Examples of the deployment of $N' = 4$ PBs to power $N = 32$ EH devices using AA–SS. The PBs deployment is based on (a) min max pairing (top), (b) min mean pairing (middle-top), and (c) cluster splitting (middle-bottom). We show in (d) how the deployment would look in the case of PBs with omnidirectional radiation patterns (e.g., using single antenna or SA, AA–IS, RPS-EMW) (bottom).

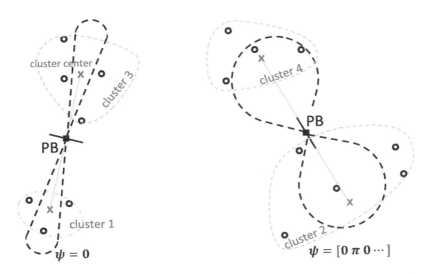

Figure 5.12 Illustration of the deployment of $N' = 2$ PBs (one using $\psi = 0$ and the other using ψ according to (4.35)) in a setup with $N = 16$ EH devices.

5.3.3 RAB

As discussed in Section 4.3.2, a rotor-equipped PB using AA–SS with ψ according to (4.35) provides an omnidirectional radiation pattern (as long as the PB rotation is sufficiently smooth, for instance using stepper rotor with at least M steps) [227]. Therefore, if all the PBs to be deployed share the same number of transmit antennas M, the deployment can be executed as in case of single-antenna PBs (Sections 5.1.1 and 5.1.2). However, when the number of transmit antennas per PB is different, clusters enclosing devices more distant from the center should be assigned PBs with larger numbers of transmit antennas[3]. Moreover, the power control must be tuned to consider the transmit gains coming from the use of the RAB strategy, which match $0.85\sqrt{M}$ for LOS channels[4]. Note that E_i in (5.5b) would then become

$$E_i \approx 0.85 \sum_{j=1}^{N'} \sqrt{M_j} p_j \varrho_{i,j}, \tag{5.9}$$

where M_j is the number of transmit antennas of T_j; while the sub-optimal power allocation given in (5.6) becomes

---

3 Note that this does not hold in general when using the native (un-rotational) AA–SS approach since in such cases the energy beams become narrower as M increases.

4 This is obtained from evaluating (4.33) with $\kappa \to \infty$, which yields $\mathbb{E}[E_{\text{aa-ss}}] = \beta(\psi)/M$, and using (4.40).

$$p_j^{\text{opt}} \approx \frac{M_j^{-1/2} \max_i(\varrho_{i,j}^{-1})}{\sum_{j=1}^{N'} M_j^{-1/2} \max_i(\varrho_{i,j}^{-1})}. \tag{5.10}$$

5.3.4 Positioning-aware CSI-free Schemes

Note that the CSI-free EB design of the previously considered schemes do not exploit a device's positioning information. Positioning information is only considered to set the ULA orientation in cases of AA–SS strategies, and power control in case of the rotary AA–SS implementation in PBs with different numbers of transmit antennas. However, greater gains are expected from combining the latter with future positioning-aware CSI-free WET schemes as envisioned in Section 4.3.3. Herein, we do not delve further into such scenarios.

5.3.5 Numerical Examples

In the following, we illustrate the system performance under the deployment strategies discussed in previous subsections. We assume each PB being equipped with a ULA, while channels undergo Rician fading with LOS factor of 10 dB. First, the average worst-case RF energy available at a set of $N = 64$ EH devices being wirelessly powered by a varying number N' of PBs is shown in Figure 5.13(a). Note that the system performance under all the CSI-free strategies providing an omnidirectional radiation pattern improves as the number of PBs increases, especially when using the RAB scheme, which provides a gain of $\sim 0.85 \times \sqrt{M}$ with respect to SA, AA–IS, RPS-EMW and single antenna WET schemes. This is no longer the situation when AA–SS combined with cluster pairing (which only appears for $N' \leq 4$ since it becomes extremely computationally expensive for greater N') and cluster splitting strategies are used. This is because as N' grows larger, it becomes increasingly difficult to set the radiation pattern to efficiently match the spatial distribution of the IoT devices. However, note that for $4 \leq N' \leq 16$, these strategies outperform the un-rotational omnidirectional radiation-based schemes for the illustrated scenario. On the other hand, Figure 5.13(b) shows the performance as a function of the number of EH devices N when $N' = 8$ PBs, with either $M = 2$ or $M = 6$ antennas, are deployed. Once again, the RAB scheme is the clear winner, but note that important gains are also reachable by using AA–SS with cluster splitting even when rotors are not incorporated in the PBs.

In general, omnidirectional radiation patterns tend to be preferable as the number of devices increases. Therefore, certain clusters where devices are more angularly dispersed may benefit from having less active antennas while using AA–SS with cluster splitting. Hence, further performance gains are expected from optimizing the number of active antennas that would be necessary at each PB. Low-complexity mechanisms addressing the combinatorial complexity of such problems are particularly attractive and deserve further research.

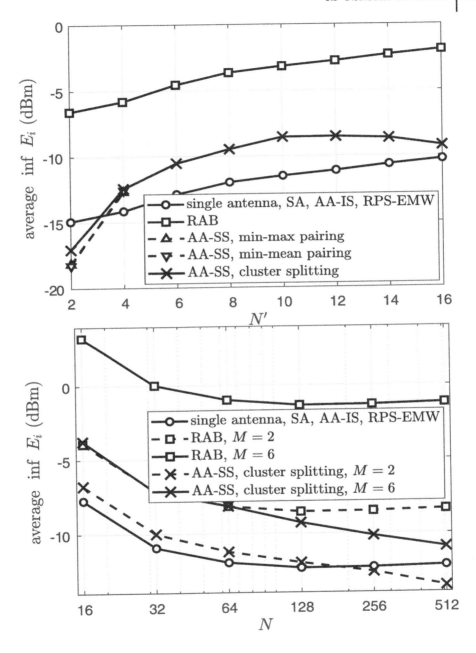

Figure 5.13 Average worst-case RF energy available at a set of *N* EH devices powered by *N'* PBs. Performance as a function of (a) the number of deployed PBs *N'*, where each PB is equipped with *M* = 4 antennas, and *N* = 64 (top), and (b) the number of deployed EH devices for *N'* = 8 and *M* ∈ {2,6} (bottom).

5.4 On the Deployment Costs

Let us define the monetary cost $C \in \mathbb{R}^+$ of the PBs deployment as

$$C = \sum_{j=1}^{N'} q_j, \tag{5.11}$$

where q_j is the monetary cost of acquiring and deploying a certain PB T_j. We assume that all the PBs to be deployed share the same features, thus, $q_j = q$, $\forall j$, and $C = qN'$. In general, q depends on the hardware and software features of the PB, but herein we are just interested on its connection to the number of PB antennas M. In the following, we consider a linear and logarithmic cost q as a function of M as follows

$$q^n(M) = \zeta_1 M, \tag{5.12}$$

$$q^g(M) = \zeta_1 + \zeta_2 \ln(M). \tag{5.13}$$

Note that ζ_1 is the nominal cost of a single-antenna PB in all cases. Figure 5.14 illustrates the cost curves that result from the above functions for $\zeta_1 = 1$ and

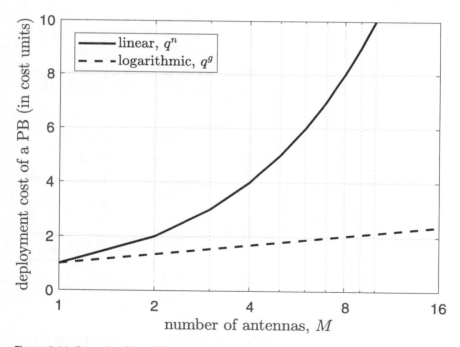

Figure 5.14 Example of the deployment cost of a PB as a function of the number of antennas M when using the linear and logarithmic cost functions given in (5.12) and (5.13), respectively. We set $\zeta_1 = 1$ and $\zeta_2 = 0.5$.

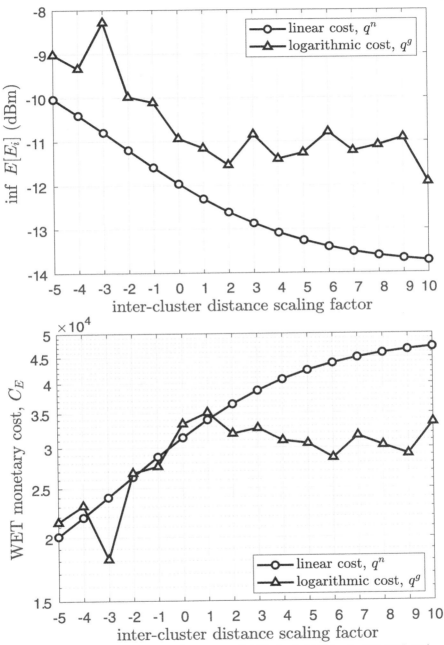

Figure 5.15 (a) Minimum average RF energy availability for $C \leq 2.8$ (cost units) (top), and (b) associated WET monetary cost (bottom), as a function of the inter-cluster distance scaling factor. Herein, we consider the example scenario in Figure 5.1 (using K-Chebyshev) with a 30 m × 15 m area as reference, i.e., 0 dB inter-cluster scaling factor, and centralized EB.

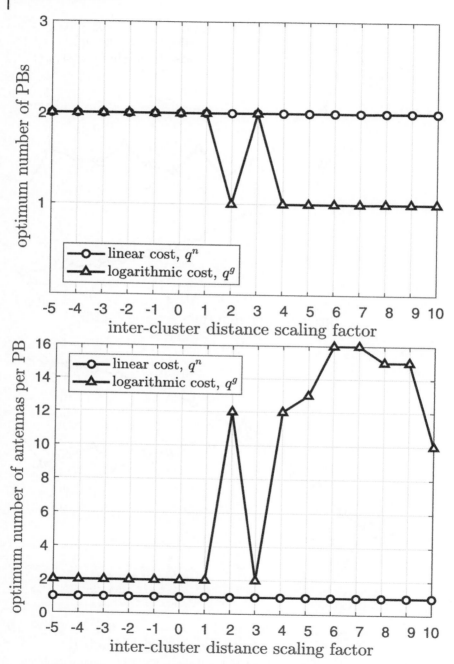

Figure 5.16 (a) Optimum number of PBs, *N'* (top), and (b) optimum number of per-PB antennas, *M* (bottom) for the results illustrated in Figure 5.15.

ζ_2 = 0.5. For the purpose of illustrating some insights, consider each PB is equipped with a ULA (in the case $M \geq 2$) and channels undergo Rician fading with LOS factor of 10 dB. When PBs are equipped with multiple antennas, we assume they use centralized EB.

First, consider the numerical results in Figure 5.15. Specifically, Figure 5.15(a) shows the max min RF energy that can be delivered to the network with a budget cost constraint set to $C \leq 2.8$ (cost units). Note that when the total cost C is constrained, q^g allows the realization of a larger set of feasible (N', M) combinations with respect to q^n, thus, it is expected that the former outperforms the latter in terms of provided energy guarantees. As observed in the figure, the energy availability guarantees worsen as the area enlarges. Meanwhile, Figure 5.15(b) depicts the associated WET monetary cost, C_E, of previous configurations. The WET monetary cost is defined as the cost of delivering a Watt unit to the worst performing device in the network, i.e.,

$$
C_E = \frac{C}{\inf_{i=1,\cdots,N} \mathbb{E}\left[E_i\right]}. \tag{5.14}
$$

In principle, the cost should steadily increase as the area gets larger, which is a phenomenon exactly observed for the case of linearly increasing antenna costs, q^n. However, in the case of logarithmic antenna costs, q^g, it may become more economic to properly adjust the number of PBs and per-PB antennas as the network changes its dimensions as illustrated in Figure 5.16. Note that the optimum configuration in the case of linear antenna costs is N' = 2 and M = 1 independently of the network dimensions, while it may be preferable to trade a PB for additional transmit antennas in case of logarithmically increasing antenna costs as the network dimensions increase.

5.5 Final Remarks

In this chapter, we considered the deployment of multiple PBs to charge a large set of EH IoT devices in a relatively wide network area. Specifically, we discussed some optimization approaches for an efficient deployment of the PBs, which either relied on the IoT device's positioning information or were completely agnostic of such information. For the case when device positioning is available during deployment of PBs, we considered K-Means clustering algorithms, and modified the way the cluster centroids are computed to improve the system performance in terms of worst-case RF energy availability. On the other hand, we proposed a greedy iterative algorithm that outputs near-optimum PBs deployment when the device's positioning information is not available at the network side and/or the devices are massively and homogeneously distributed in the network.

Additionally, we described how the transmit EB can be designed in the case of deploying multi-antenna PBs by re-using some of the results of Chapter 4. Specifically, we analyzed centralized EB, in which each PB acquires the CSI of the channels toward all the devices in the system, and distributed EB, in which each PB exploits only the CSI of the channels toward its surrounding/associated EH devices. We illustrated performance comparisons between them, which turned out to favor the former. However, the considered centralized EB is ideal in the sense that it assumes perfect, instantaneous, and cost-free CSI-acquisition, and high-capacity control channels between the PBs and a central unit that runs the optimization. How the performance gap would vary if any of such ideal conditions do not hold remains an open problem.

Note that the PBs deployment optimization and EB design are in general two problems that cannot be coupled since the latter commonly requires some kind of time-varying CSI. However, this does not necessarily hold in cases where the multi-antenna PBs are deployed to operate without CSI but still get some spatial gains, as in the case of implementing the CSI-free schemes discussed in Section 4.3. For such scenarios, we showed that the deployment can still be carried out as in the case of single-antenna PBs, when SA, AA–IS, RPS-EMW or RAB schemes are to be used by the PBs. Meanwhile, we proposed novel heuristic optimization methods for the deployment of PBs operating with the AA–SS CSI-free scheme. The advantages and disadvantages of each optimization approach were also discussed.

Finally, we discussed some numerical results related to the PBs deployment costs. This was done by considering PBs acquisition costs increasing linearly or logarithmically with the number of equipped antennas. We found that the former favors the acquisition and deployment of single-antenna PBs, while the latter may favor the deployment of multi-antenna PBs, specially as the network area widens.

6

Wireless-powered Communication Networks

6.1 WPCN Models

In a WPCN, energy transmitters perform WET to EH devices in a first phase, then, the devices use the harvested energy to transmit their own data to information APs in a second phase. Figure 6.1 illustrates the two basic WPCN models: 1. separate energy transmitter (dedicated PB) and information receiver, and 2. co-located energy transmitter and information receiver (hybrid PBs (HPBs) or equivalently hybrid APs(HAPs)[1]). The model with separate PBs and APs (Figure 6.1(a)) makes the coordination of information and energy transmissions more challenging, and production, deployment and operation costs greater compared to the integrated HAP model (Figure 6.1(b)). However, such co-located WPCN also induces a practical design challenge named the doubly-near-far problem [162], where a user that is far away from its associated HAP harvests less energy in the WET phase but may consume more to transmit data in the WIT phase than that of a user nearer to the HAP. As a result, unfair user performance may occur since a distant user's throughput can be much smaller than that of a nearby user.

The user unfairness coming from the doubly-near-far problem can be alleviated by using intelligent resource allocation and cooperation schemes, moving (motor-equipped) HAPs, and multi-antenna nodes. Note that the less energy constrained IoT devices can assist (via information relaying) the information transmissions from the more energy constrained devices. Meanwhile, the EH devices can experience different performance over time when they and/or the HAPs change their relative positions. In case of moving HAPs, the average performance can be optimized by properly tuning the moving trajectories. Finally, the transmit energy/receive information precoders/combiners can be designed to favor the farthest devices in the WET, WIT, or both phases when HAPs are equipped with multiple-antennas.

---

1 Since the terms HPB and HAP are equivalent, hereafter we merely use the latter.

Wireless RF Energy Transfer in the Massive IoT Era: Towards Sustainable Zero-energy Networks.
First Edition. Onel Alcaraz López and Hirley Alves.
© 2022 by The Institute of Electrical and Electronics Engineers, Inc. Published 2022 by John Wiley & Sons, Inc. Companion website: www.wiley.com/go/Alves/WirelessEnergyTransfer

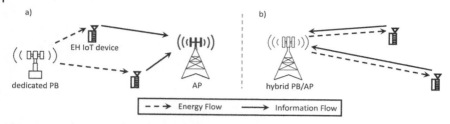

Figure 6.1 Basic models of WPCNs: (a) separated PBs and APs (left), and (b) collocated PBs and APs (right).

Moreover, WET and WIT can be performed either in an out-of-band (over different frequency resources) or in-band (over the same frequency resources) manner [78]. The former implementation is easier to design, and interference to nearby networks can be easily controlled; although spectrum efficiency may be seriously affected, and serving EH devices with different RF systems becomes challenging. Meanwhile, the latter is preferable in terms of spectral efficiency when heterogeneous devices coexist, and the interference from WET signals to nearby/coexisting information networks can be mitigated by using adequate precoders and/or deterministic (known by the network) WET signals, which the ID devices can mitigate using SIC. Note that synchronization/coordination between the PBs, APs and HAPs may be needed in order to maintain the associated devices with updated information of the network WET signals. In general, integrating two set of antennas in the IoT devices/APs (one for RF-EH/WET and the other for ID/WIT) may be necessary for optimum performance since energy and information transceivers normally require different antenna and RF systems.

Additionally, novel WET and/or WIT full-duplex (FD) schemes may boost the performance of WPCNs. An EH IoT device operating in a FD manner (with at least one antenna for RF–EH and another for WIT) is able to transmit information and harvest energy to/from the AP/PB (HAP) in the same band. For instance, when the PB and AP in Figure 6.1(a) are well separated such that an energy transmission does not cause strong interference to ID, the devices may simultaneously receive energy from the PB and transmit information to the AP. This enables an additional benefit known as self-energy recycling [157] since an extra RF energy can be harvested from the device's own transmitted information signal. Finally, another promising solution for an in-band approach is to use FD HAPs that are able to transmit energy and receive information to/from the IoT devices simultaneously in the same frequency band. Although many studies suggest that accurate SIC is difficult to achieve, an FD HAP has greater potential to provide folded spectrum efficiency improvement than conventional half-duplex energy/information transmissions.

6.2   Reliable Single-user WPCN

As discussed in Chapter 1, many IoT use cases towards the 6G era may require extreme per-packet reliability guarantees. This, combined with ultra-low energy availability and short message transmission due to intrinsically small data payload and/or ultra-low latency requirements, make the design of the physical layer of reliable WPCNs extremely challenging. In this section, we focus our attention on the reliability performance of a single-user WPCN, while we also evaluate the user's energy consumption.

We consider a collocated WPCN as in Figure 6.1(b) with a single-antenna HAP and IoT node S; however, the analysis and proposed protocols can be easily extended to a WPCN with a multi-antenna HAP or a separate PB and AP as in Figure 6.1(a). Moreover, S is equipped with a finite battery of capacity B_{max}. In a first phase, the HAP charges S during l channel uses (WET phase), and doing this acts as an interrogator, requesting information from S. Then, S transmits b information bits over n channel uses in a WIT phase. We define a "transmission round" as a pair of consecutive WET and WIT phases, in that order. In general, S can transmit its data using all the energy available in its battery at the start of each WIT phase (without energy accumulation between transmission rounds) or just make use of a part of such energy, saving the rest for future transmissions (with energy accumulation between transmission rounds). We consider a time-constrained setup, which implies that the HAP has to decode the received signal for each arriving information block, which in turn is of short block length, for instance, relatively small n.

Channels are assumed to undergo Nakagami-m fading, which can model different fading environments by adjusting its parameters. Moreover, WET and WIT phases are scheduled to use the same frequency resources, thus we assume the corresponding channels are reciprocal. The path loss and normalized channel gains are denoted by ϱ and z, respectively, where $z \sim \Gamma(m, 1/m)$. Finally, the duration of a channel use is denoted by T_c, and CSI is assumed at the HAP for data decoding.

6.2.1   Harvest-then-transmit (HTT)

Herein, we discuss and analyze the performance of the HTT protocol [162], under which all the energy harvested in the WET phase is used for transmission in the subsequent WIT phase.

WET Phase. In this phase, the HAP charges S during l channel uses. Herein, we adopt the "above sensitivity" EH transfer model [76] (see Section 2.2), thus, the energy harvested at S during the tth transmission round can be expressed as

$$E^{h^{(t)}} = \begin{cases} 0, & \text{if } P\varrho z^{(t)} < \varpi_1 \\ \eta P\varrho z^{(t)} l T_c, & \text{if } P\varrho z^{(t)} \geq \varpi_1 \end{cases}$$

$$= \mathbb{I}(z^{(t)} \geq z_1)\eta P\varrho z^{(t)} lT_c \tag{6.1}$$

where $\eta \in (0,1)$ is a constant PCE, P is the transmit power of the HAP, and ϖ_1 is the sensitivity RF level of the EH circuit at the IoT device, thus $z_1 = \varpi_1/(P\varrho)$ is the corresponding channel sensitivity threshold. Note that the energy harvested from noise has been neglected since in practice its contribution is extremely small.

Harvested energy is first stored in a rechargeable battery of capacity $B_{\max}$, and becomes available in the current round. Then, the charge of the battery at the beginning of the tth WIT phase is updated as follows

$$
\begin{aligned}
B^{(t)} &= \min(B_{\max}, E^{h^{(t)}}) = \min\left(B_{\max}, \mathbb{I}(z^{(t)} \geq z_1)\eta P\varrho z^{(t)} lT_c\right) \\
&= \mathbb{I}(z^{(t)} \geq z_1)\min\left(B_{\max}, \eta P\varrho z^{(t)} lT_c\right) \\
&= \mathbb{I}(z^{(t)} \geq z_1)\min(z^{(t)}, z_2)\eta P\varrho lT_c,
\end{aligned}
\tag{6.2}
$$

where $z_2 = B_{\max}/(\eta P\varrho lT_c)$ is the channel power gain threshold for the saturation of the battery in S.

WIT Phase. In the WIT phase, S uses all the energy (if any) in its battery to transmit a message of b bits to the HAP over n channel uses. Thus, the transmit power in the tth round is $p^{(t)} = B^{(t)}/(nT_c)$, and the instantaneous SNR at the HAP can be written as

$$\gamma^{(t)} = \frac{\varrho p^{(t)} z^{(t)}}{\Upsilon\sigma^2} = \frac{\varrho B^{(t)} z^{(t)}}{nT_c\Upsilon\sigma^2} = \mathbb{I}(z^{(t)} \geq z_1)\min(z^{(t)}, z_2)z^{(t)}\beta, \tag{6.3}$$

where Υ represents the SNR gap from the additive white Gaussian noise (AWGN) channel capacity due to a practical modulation and coding scheme being used, and $\beta = \frac{\eta P\varrho^2 l}{n\Upsilon\sigma^2}$. From (6.3), observe that $\gamma^{(t)}$ is proportional to the square of the power channel coefficient as long as the battery of S is not saturated, otherwise, it only relies on the scaled power channel coefficient.

Error Probability. The decoding error probability ε at the HAP is not only a function of the receive SNR and data rate, but also of the information block length n, specially when n is relatively small. For more details about finite block length communication see Appendix A. Regarding this, a tight approximation for $\varepsilon(\gamma, b, n)$ for AWGN channels is given in (A.7). However, there is often no practical need of making use of such a finite block length-based framework to evaluate the average decoding error probability in the presence of quasi-static fading channels, $\mathbb{E}_\gamma[\varepsilon(\gamma, b, n)]$. This is because such performance metric approximates the asymptotic outage probability, $\mathbb{P}[\gamma \leq 2^{b/n} - 1]$, in such scenarios when b/n is not extremely small and there is not a very strong LOS component [77, 236, 237]. Herein, we adopt such approximation to facilitate the analytical derivations. Let us proceed as follows. Note that (6.3) can be rewritten as

$$\gamma^{(t)} = \begin{cases} 0, & z^{(t)} \leq z_1 \\ \beta(z^{(t)})^2, & z_1 < z^{(t)} < z_2 \\ \beta z_2 z^{(t)}, & z^{(t)} \geq z_2 \end{cases} \tag{6.4}$$

for $z_2 \geq z_1$, which must always hold as otherwise EH can never occur. Now, the outage probability can be computed as

$$P_{\text{out}}^{\text{htt}} = \mathbb{P}\left[\gamma < 2^{\frac{b}{n}} - 1\right]$$

$$= \mathbb{P}\left[z \leq z_1 \bigcup \left(\beta z^2 < 2^{\frac{b}{n}} - 1 \bigcap z_1 < z < z_2\right)\right.$$

$$\left.\bigcup \left(\beta z_2 z < 2^{\frac{b}{n}} - 1 \bigcap z \geq z_2\right)\right]$$

$$= \mathbb{P}\left[z \leq \max\left(z_1, \min\left(z_2, \sqrt{\frac{2^{b/n}-1}{\beta}}\right)\right) \bigcup\right.$$

$$\left.\max\left(z_2, \sqrt{\frac{2^{b/n}-1}{\beta}}\right) \leq z \leq \max\left(z_2, \frac{2^{b/n}-1}{\beta z_2}, \sqrt{\frac{2^{b/n}-1}{\beta}}\right)\right]$$

$$= F_Z\left(\max\left(z_1, \min\left(z_2, \sqrt{\frac{2^{b/n}-1}{\beta}}\right)\right)\right)$$

$$+ F_Z\left(\max\left(z_2, \frac{2^{b/n}-1}{\beta z_2}, \sqrt{\frac{2^{b/n}-1}{\beta}}\right)\right) - F_Z\left(\max\left(z_2, \sqrt{\frac{2^{b/n}-1}{\beta}}\right)\right), \tag{6.5}$$

where the CDF of Z is $F_Z(z) = 1 - \frac{\Gamma(m,mz)}{\Gamma(m)}$. On the other hand, (6.5) can be rewritten for relatively large batteries as

$$\lim_{z_2 \to \infty} P_{\text{out}}^{\text{htt}} = \mathbb{P}\left[z \leq \max\left(z_1, \sqrt{\frac{2^{b/n}-1}{\beta}}\right)\right] = F_Z\left(\max\left(z_1, \sqrt{\frac{2^{b/n}-1}{\beta}}\right)\right). \tag{6.6}$$

Average Power Consumption. Herein, we mathematically characterize the IoT device's average power consumption, i.e., average transmit power, under the HTT protocol operation. We proceed as follows

$$\bar{P}_{\text{cons}}^{\text{htt}} = \mathbb{E}_t\left[p^{(t)}\right] = \mathbb{E}_t\left[\frac{B^{(t)}}{nT_c}\right]$$

$$\stackrel{(a)}{=} \mathbb{E}_z\left[\mathbb{I}(z \geq z_1)\min(z, z_2)\right]\eta P\varrho l/n$$

$$= (\eta P \varrho l/n) \int_0^\infty \mathbb{I}(z \geq z_1) \min(z, z_2) f_Z(z) dz$$

$$\overset{(b)}{=} (\eta P \varrho l/n) \left(\int_{z_1}^{z_2} z f_Z(z) dz + z_2 \int_{z_2}^\infty f_Z(z) dz \right)$$

$$\overset{(c)}{=} (\eta P \varrho l/n) \left(\frac{m^m}{\Gamma(m)} \int_{z_1}^{z_2} z^m e^{-mz} dz + z_2(1 - F_Z(z)) \right)$$

$$\overset{(d)}{=} (\eta P \varrho l/n) \left(\frac{\Gamma(m+1, mz_1) - \Gamma(m+1, mz_2)}{\Gamma(m+1)} + z_2 \frac{\Gamma(m, mz_2)}{\Gamma(m)} \right), \tag{6.7}$$

where (a) comes from using (6.2) and from transforming the expectation over time (transmission rounds) to the expectation over the fading realizations. Meanwhile, (b) follows after separating the integrating intervals and using $z_2 \geq z_1$, which needs to hold in practice. Then, (c) comes from using the PDF of Z, $f_Z(z) = m^m z^{m-1} e^{mz} / \Gamma(m)$, and using the definition of the CDF. Finally, we attain (d) after some algebraic transformations of the incomplete gamma function definition [238, Eq.(8.2.1)].

For relatively large batteries, (6.7) can be simplified to obtain

$$\lim_{z_2 \to \infty} \bar{P}_{\text{cons}}^{\text{htt}} = (\eta P \varrho l/n) \int_{z_1}^\infty z f_Z(z) dz = (\eta P \varrho l/n) \frac{\Gamma(m+1, mz_1)}{\Gamma(m+1)}. \tag{6.8}$$

Interestingly, when $z_1 = 0$ in (6.7), and for scenarios in which the fading is less severe, for instance, larger m, the average power consumption increases asymptotically approaching the case of infinite battery (6.8) for practical systems where $z_2 > 1$. This is

$$\lim_{m \to \infty} \bar{P}_{\text{cons}}^{\text{htt}} = (\eta P \varrho l/n) \min(z_2, 1), \tag{6.9}$$

which makes sense since the channel tends to behave like an AWGN channel and no battery saturation occurs for $z_2 > 1$.

Finally, the average energy consumption can be computed as $n T_c \bar{P}_{\text{cons}}^{\text{htt}}$.

6.2.2 Allowing Energy Accumulation

Herein, we analyze the scenario where the EH device is allowed to save energy for future transmission rounds. The battery state evolution is given by

$$B^{(t)} = \min(B_{\max}, B^{(t-1)} + E^{h^{(t)}} - p^{(t-1)} n T_c), \tag{6.10}$$

with $E^{h^{(t)}}$ obeying (6.1).

Next, we discuss the power control mechanism proposed in [76] that leverages CSI at S in terms of channel power gain (thus, no specific channel training is needed since this information can be acquired from the instantaneous harvested energy) to improve the system performance compared to that offered by the HTT protocol. The protocol itself is independent from the channel fading distribution, not being restricted to Nakagami-m fading, which is only used later to illustrate some numerical performance figures.

Fixed-error inversion power control (FEIPC). The idea behind FEIPC[2] is as follows. At the tth transmission round, and given a known target SNR at the HAP, γ_{th}, and the instantaneous channel gain,

- S determines whether it has sufficient energy in its battery to transmit with such a power $p^{(t)}$ that allows the information signal to be sent to reach the HAP with SNR γ_{th};
- if there is sufficient energy, S transmits with $p^{(t)}$, otherwise, it stays silent and saves energy for future WIT rounds.

Thus,

$$p^{(t)} = \begin{cases} \dfrac{\gamma_{th}\Upsilon\sigma^2}{\varrho z^{(t)}}, & \text{if } \dfrac{\gamma_{th}\Upsilon\sigma^2}{\varrho z^{(t)}} \le \dfrac{B^{(t)}}{nT_c} \\ 0, & \text{otherwise} \end{cases}. \tag{6.11}$$

Optimum SNR Threshold. Now, the question that arises is: *what is the minimum SNR threshold γ_{th} for which the system operates with maximum reliability?* In case of favorable EH conditions and considering sufficiently long data messages such that Shannon capacity is attainable, the answer is simple: $\gamma_{th}^{opt} = 2^{b/n} - 1$, since this is the minimum possible for error-free communication. Such asymptotic design was first proposed by Isikman *et al.* in [239] for a general (not necessarily RF) EH setup. However, in practice, sudden unfavorable channel conditions, which hinder the EH process and lead S to transmit with higher power, may prevent some WIT rounds executing, for instance $p^{(t)} = 0$; while the transmission of finite blocklength messages makes it impossible to reach a decoding error-free operation region. In such a scenario, the overall error probability is given by

$$\varepsilon_{overall}(\gamma_{th}) = \left(1 - \varepsilon_{out}(\gamma_{th})\right)\varepsilon(\gamma_{th}, b, n) + \varepsilon_{out}(\gamma_{th}), \tag{6.12}$$

where

$$\varepsilon_{out}(\gamma_{th}) = \mathbb{P}[p^{(t)} = 0] = \mathbb{P}\left[\frac{\gamma_{th}\Upsilon\sigma^2}{\varrho z^{(t)}} \le \frac{B^{(t)}}{nT_c}\right] \tag{6.13}$$

is the probability that the energy available in the battery is insufficient for achieving the required γ_{th} at the HAP. Meanwhile $\varepsilon(\gamma_{th}, b, n)$ represents the

---

2 In [76], it is referred as finite block length fixed threshold transmission (FB-FTT).

decoding error, which happens to equal that of a message transmitted over n channel uses of an AWGN channel with SNR γ_{th} with rate b/n bits per channel use (bpcu), thus given by (A.7).

For a given decoding error probability target ε_0, we can obtain the following result regarding the corresponding SNR threshold γ_{th}.

Theorem 6.1 [76, Lemma 1] The required γ_{th} to reach any ε_0 is a unique solution to

$$\gamma_{th} = 2^{\frac{b}{n} + \frac{1}{\sqrt{n}} \mathcal{V}(\gamma_{th}) Q^{-1}(\varepsilon_0)} - 1, \tag{6.14}$$

where $\mathcal{V}(x) = \sqrt{1 - \frac{1}{(1+x)^2}}$. Algorithm 3 converges to find the solution of (6.14) for $\varepsilon_0 \leq 0.5$.

Proof. In Appendix D, we reproduce the proof provided in [76] adapted to our notation here. Therein, we also provide evidence of the fast convergence of Algorithm 3 in practical setups. □

Now, note that the higher the value of ε_0 is, the smaller γ_{th} and transmit power p are required, and the smaller the value of ε_{out} becomes, and vice versa. Unfortunately, a closed-form expression for ε_{out} seems intractable, and we later resort to simulations in order to compute it. In addition, we can see that $\varepsilon_{overall} \geq \varepsilon_0$ always holds, thus a relatively high value of ε_0 can seriously limit the system performance for some setups. In fact, numerical evidence provided in [76, App. D] suggests that there is a unique optimum value of ε_0, ε_0^{opt}, that minimizes the overall error probability in practical setups.

Algorithm 3 Iterative solution of (6.14)

1: **Input:** ε_0, b, n
2: Initialize $\mathcal{V}(\gamma_{th}) = 1$
3: **repeat**
4:   Update γ_{th} by evaluating (6.14)
5:   Update $\mathcal{V}(\gamma_{th})$, i.e., $V(\gamma_{th}) \leftarrow \sqrt{1 - \frac{1}{(1+\gamma_{th})^2}}$
6: **until** convergence
7: **Output:** γ_{th}

Average Power Consumption. The average transmit power of S when using the aforementioned power control scheme is given by

$$\bar{P}_{cons}^{feipc}(\gamma_{th}) = \mathbb{E}_t[p^{(t)}] \overset{(a)}{=} (1 - \varepsilon_{out}(\gamma_{th})) \int_0^{z_2} \frac{\gamma_{th} \Upsilon \sigma^2}{\varrho z} f_Z(z) dz$$

$$\overset{(b)}{=} (1 - \varepsilon_{\text{out}}(\gamma_{\text{th}})) \int_0^{z_2} \frac{\gamma_{\text{th}} \Upsilon \sigma^2 m^m}{\varrho \Gamma(m)} z^{m-2} e^{-mz} dz$$

$$\overset{(c)}{=} -(1 - \varepsilon_{\text{out}}(\gamma_{\text{th}})) \frac{\gamma_{\text{th}} \Upsilon \sigma^2 m}{\varrho \Gamma(m)} \Gamma(m-1, mz) \Big|_0^{z_2}$$

$$= (1 - \varepsilon_{\text{out}}(\gamma_{\text{th}})) \gamma_{\text{th}} \Upsilon \sigma^2 m \left(\frac{1}{m-1} - \frac{\Gamma(m-1, mz_2)}{\Gamma(m)} \right) / \varrho$$

$$\tag{6.15}$$

for $m > 1$, which is the case of more practical interest. Note that (a) and (b) come from using (6.11) and the PDF of Z, respectively, while (c) follows from algebraic transformations of the incomplete gamma function definition with $m > 1$ [238, Eq.(8.2.1)].

Now, for relatively large batteries, (6.15) can be simplified to obtain

$$\lim_{z_2 \to \infty} \bar{P}_{\text{cons}}^{\text{feipc}}(\gamma_{\text{th}}) = -(1 - \varepsilon_{\text{out}}(\gamma_{\text{th}})) \frac{\gamma_{\text{th}} \Upsilon \sigma^2 m}{\varrho \Gamma(m)} \Gamma(m-1, mz) \Big|_0^\infty$$

$$= (1 - \varepsilon_{\text{out}}(\gamma_{\text{th}})) \frac{\gamma_{\text{th}} \Upsilon \sigma^2 m}{\varrho(m-1)}. \tag{6.16}$$

Note that when the fading is less severe, for example, larger m, ε_{out} decreases while the remaining terms depending on m tend to unity. This is because asymptotically, and considering $z_1 < 1$, which has to be true in practice, we have that

$$\lim_{m \to \infty} \bar{P}_{\text{cons}}^{\text{feipc}}(\gamma_{\text{th}}) = \gamma_{\text{th}} \Upsilon \sigma^2 \mathbb{I}(z_2 > 1)/\varrho, \tag{6.17}$$

$$\lim_{m \to \infty} \lim_{z_2 \to \infty} \bar{P}_{\text{cons}}^{\text{feipc}}(\gamma_{\text{th}}) = \gamma_{\text{th}} \Upsilon \sigma^2 /\varrho. \tag{6.18}$$

Therefore, an average transmit power very close to $\gamma_{\text{th}} \Upsilon \sigma^2$ is expected for any practical system with $m \gg 1$. In fact, the instantaneous transmit power tends to be exactly $\gamma_{\text{th}} \Upsilon \sigma^2$ since the channel tends to behave as an AWGN channel at the same time that no saturation or complete depletion of the battery ever occurs.

Performance Gap with Respect to the Asymptotic Analysis. The gap between the finite block length and asymptotic SNR threshold can be computed as follows

$$\delta = \frac{\gamma_{\text{th}}}{2^{b/n} - 1} \overset{(a)}{=} \frac{2^{\frac{b}{n} + \frac{1}{\sqrt{n}} \mathcal{V}(\gamma_{\text{th}}) \log_2 e \, Q^{-1}(\varepsilon_0)} - 1}{2^{b/n} - 1}$$

$$\overset{(b)}{=} e^{\frac{1}{\sqrt{n}} Q^{-1}(\varepsilon_0)} + \frac{e^{\frac{1}{\sqrt{n}} \mathcal{V}(\gamma_{\text{th}}) \, Q^{-1}(\varepsilon_0)} - 1}{2^{b/n} - 1}, \tag{6.19}$$

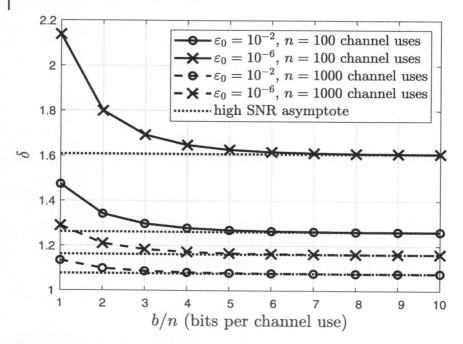

Figure 6.2 δ as a function of the transmit rate for $n \in \{100,1000\}$ channel uses and $\varepsilon_0 \in \{10^{-2}, 10^{-6}\}$.

where (a) comes from using (6.14), although notice that $\mathcal{V}(\gamma_{th})$ is still a function of γ_{th}, while (b) comes from simple algebraic transformations. Meanwhile, when operating at high data rates, for instance, high SNR regime, (6.19) can be further simplified to attain

$$\lim_{g/n \to \infty} \delta = e^{\frac{1}{\sqrt{n}}Q^{-1}(\varepsilon_0)} \tag{6.20}$$

since $\mathcal{V}(\infty) = 1$. Moreover, since δ is a decreasing function of b/n, we conclude that (6.20) constitutes a lower bound of (6.19).

Figure 6.2 illustrates the asymptotic behavior of (6.20) for $n \in \{100, 1000\}$ channel uses and ε_0. The main point from (6.20), which is also shown in the figure, is that the required SNR has to be at least $e^{\frac{1}{\sqrt{n}}Q^{-1}(\varepsilon_0)}$ times greater than the usual threshold of $2^{b/n} - 1$ to reach an error probability no lower than ε_0 while transmitting the information through n channel uses. Obviously, this criterion is also applied to the transmit power. Additionally, notice that the asymptotic bounds illustrated in Figure 6.2 are tight already for $b/n \sim 4$ bits per channel use since $\mathcal{V} > \sqrt{1 - \frac{1}{(2^4)^2}} = 0.998$, and $2^4 - 1 = 15$ is at least 10 times greater than the numerator of the fraction in the last equality of (6.19) for any combination of $n \geq 100$ channel uses and $\varepsilon_0 \geq 3 \times 10^{-20}$. However, the lower the data rate and/or the shorter the information block length and/or the more stringent the target error probability, the greater the required SNR with respect to the threshold at infinite block length $2^{b/n} - 1$, thus showing how misleading it is to calculate the SNR thresholds and rates using any scheme based on the assumption of infinite block length.

6.2.3 HTT versus FEIPC

Herein, we present numerical evidence of the outstanding performance of FEIPC with respect to the basic HTT protocol. Unless stated otherwise, results are obtained by setting $P = 1$ W, $\eta = 0.5$, $\varpi_1 = 1$ μW, $T_c = 10$ μs, and $\varrho = 52$ dB, which may correspond to the path loss experienced at a distance of 21.6 m when using the log-distance model in (4.21) and (5.2). This configuration leads to an average receive RF power of ~ 6.3 μW (~ 3.15 μW of harvested energy) at S. Moreover, $m = 2$, $\sigma^2 = -90$ dBm, $b = 320$ bits, and $\Upsilon = 9.8$ dB (assuming that an uncoded quadrature amplitude modulation is employed [226]).

On the Optimal Decoding Error Threshold. Figure 6.3 presents the overall error probability (Figure 6.3(a)) and average transmit power (Figure 6.3(b)) as a function of ε_0 for $l = 800$, $n = 150$ channel uses and infinite battery capacity. The results in Figure 6.3(a) corroborate the existence and uniqueness of an optimum value of ε_0 for the FEIPC protocol, which in this case turns out to be $\varepsilon_0^{opt} = 10^{-6}$. Note that when operating with this decoding error threshold, energy outage events are extremely infrequent since $\varepsilon_{overall}^{opt} \approx 10^{-6}$.

Increasing ε_0 allows greater energy saving when using FEIPC since the average transmit power decreases as shown in Figure 6.3(b), however the error performance is bounded by this value.

Finally, we can note the remarkable performance gap between HTT and the power control protocol. In fact, the FEIPC scheme has the best performance for practical scenarios where $\varepsilon_0 \ll 1$. This reinforces the appropriateness of the idea behind saving energy for future transmission rounds.

On the Transmit Power and Battery Capacity. In Figure 6.4, we evaluate the impact of battery capacity, $B_{max} \in \{10^{-7}, 10^{-5}, \infty\}$ J, while comparing the performance of HTT and FEIPC protocols in terms of $\varepsilon_{overall}$ (Figure 6.4(a)) and $\bar{P}_{cons}$ (Figure 6.4(b)) as a function of P, for $\varepsilon_0 = 10^{-6}$, and $l = 800$, $n = 150$ channel uses. As shown in Figure 6.4 (top), the impact of a finite battery capacity on the error performance is insignificant for the HTT protocol since there is no energy accumulation between transmission rounds. Therefore, only when a high amount of energy is being transferred, e.g., $P > 45$ dBm, the corresponding performance gap starts to be appreciable. However, for the FEIPC scheme the situation is more delicate since the larger the battery capacity, the greater the chance to save more energy for future transmissions, thus the better the

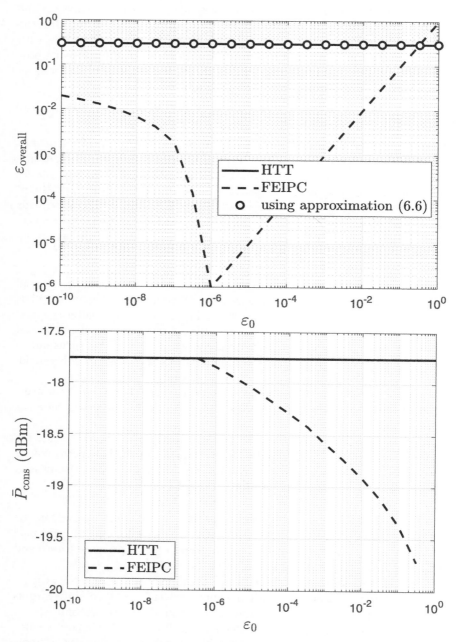

Figure 6.3 (a) $\varepsilon_{\text{overall}}$ (top) and (b) $\bar{P}_{\text{cons}}$ (bottom), as a function of ε_0 for $l = 800$ and $n = 150$ channel uses with $B_{\text{max}} = \infty$.

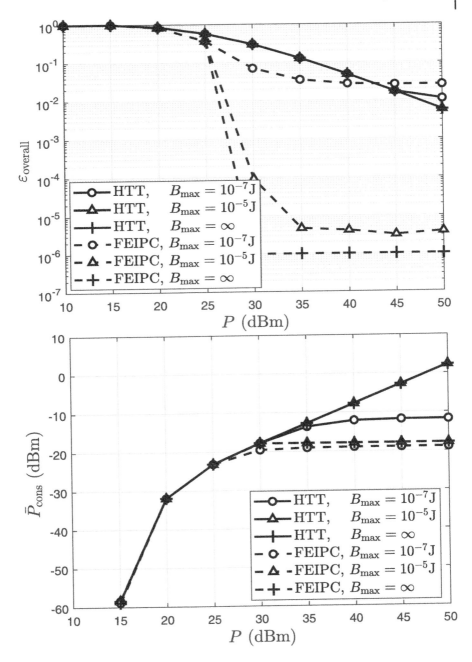

Figure 6.4 System performance as a function of P, (a) $\varepsilon_{\text{overall}}$ (top) and (b) $\bar{P}_{\text{cons}}$ (bottom), for $B_{\text{max}} \in \{10^{-7}, 10^{-5}, 10^{-3}, \infty\}$ J, $\varepsilon_0 = 10^{-6}$, and $l = 800$, $n = 150$ channel uses.

error performance (Figure 6.4(a)) and the larger the average energy consumption (Figure 6.4(b)). Notice the small system performance gap between setups with $B_{max} = 10^{-5}$ J and $B_{max} = \infty$, which is due to the small amount of energy being harvested in these setups with short WET phase.

It is evident that for $P > 30$ dBm we have $\varepsilon_0^{opt} < 10^{-6}$ since the system setup favors a better performance. Moreover, the IoT device's average transmit power under the HTT protocol remains almost constant around $\bar{P}_{cons} = -11.8$ dBm for $B_{max} = 10^{-7}$ J and $P > 40$ dBm since $\mathbb{E}[E^{h^{(t)}}] > B_{max}$, for example, $\mathbb{E}[E^{h^{(t)}}]\big|_{P=40dBm} \approx 2.5 \times 10^{-7}$ J, thus $B^{(t)} \approx B_{max}$ and $p^{(t)} = B^{(t)}/(nT_c) \approx \frac{10^{-7}J}{150 \times 10^{-5}s} = 6.7 \times 10^{-5} \rightarrow -11.8$ dBm, both holding almost all the time. The FEIPC protocol leads to an even smaller energy consumption since the IoT device does not spend all the energy available in its battery in each round, especially when channel conditions are favorable.

On the Optimum Block Length Configuration. In the results shown in Figure 6.5, we fix the system delay in delivering each message by setting $n+l = 1000$ channel uses, i.e., $1000T_c = 10$ ms. Results are given as a function of n, for $\varepsilon_0 \in \{10^{-3}, 10^{-6}, 10^{-9}\}$ and $B_{max} = 10^{-5}$ J. Notice that HTT performs poorly

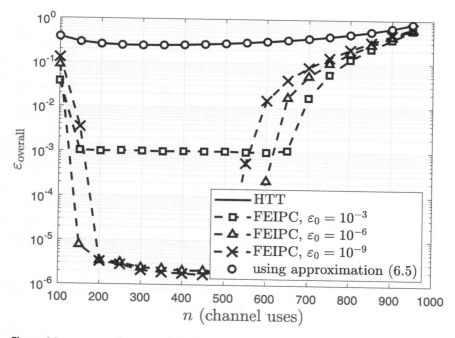

Figure 6.5 $\varepsilon_{overall}$ as a function of n for $\varepsilon_0 \in \{10^{-3}, 10^{-6}, 10^{-9}\}$, $B_{max} = 10^{-5}$ J, and fixed delay $n + l = 1000$ channel uses.

all the time, i.e., $\varepsilon_{\text{overall}} > 0.2$; while the FEIPC protocol can offer a much better reliability performance and even enable URLLC by properly setting ε_0. Observe that the FEIPC protocol achieves an error probability around $\varepsilon_{\text{overall}} = 10^{-6}$ for $n \sim 450$ channel uses. Also, when $n \geq 500$ channel uses, a target error probability greater than 10^{-6} is required in order to achieve the optimum system performance. Then, $n^{\text{opt}} \sim 450$, $l^{\text{opt}} \sim 550$ channel uses are approximately the optimum values for reaching a target error close to $\varepsilon_0^{\text{opt}} \sim 10^{-6}$ within the given delay constraint of 1000 channel uses.

All these results clearly show the convenience of a joint optimization of n, l and ε_0. In fact, an off-line optimization, which yields optimum parameters *a priori* and is valid for long periods, seems more convenient for energy-constrained setups since it avoids the interchange of additional information between the nodes. Finally, the results in Figure 6.3(a) and Figure 6.5 corroborate the accuracy of expressions (6.5) and (6.6) with respect to finite block length-based Monte Carlo simulations.

6.3 Multi-user Resource Allocation

Different from the previous scenario, here the network consists of one HAP (collocated model as in Figure 6.1(b)) equipped with M antennas, and N single-antenna EH IoT devices denoted by $\mathcal{S} = \{S_i | i = 1, 2, \cdots, N\}$. All the IoT devices store the energy harvested in the downlink WET phase in a rechargeable battery and use it to power operating circuits, and transmit training and data information in the uplink.

It is further assumed that the HAP and all the users operate over the same frequency band. Thus, uplink and downlink channels are reciprocal and denoted by a common normalized complex channel vector $\mathbf{h}_i$ when established between the HAP's antennas and user S_i. Meanwhile, ϱ_i denotes the large-scale power channel gain. Moreover, channels are subject to quasi-static flat-fading, where $\mathbf{h}_i$ remain constant during each block transmission time, but can vary from one block to another.

As illustrated in Figure 6.6, each transmission block consists of

- An overhead sub-block (of fixed time), where all the IoT devices transmit their unique pilot sequences (preferably orthogonal among one another) for CSI estimation at the HAP. Moreover, a subset of the devices, those with non-empty buffer (hereinafter referred to as the active devices), transmits a request for data transmission in the uplink WIT sub-block. After the HAP has processed all this, it proceeds to schedule uplink WIT resources for the active nodes, which are then immediately informed.
- A downlink WET sub-block of duration $(1 - \tau_0)T$, $0 \leq \tau_0 < 1$, where the HAP transmits energy signals to power the entire set of devices according to their energy needs.

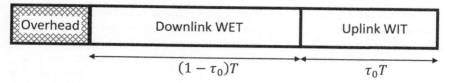

Figure 6.6 WET and WIT scheduling protocol.

- An uplink WIT sub-block of duration $\tau_0 T$, where the active devices transmit their data messages according to the resource scheduling executed, and previously notified, by the HAP.

We assume that the HAP acquires a perfect CSI of all the channels after the overhead sub-block processing. However, hereinafter we neglect the overhead sub-block, and focus only on the WET and WIT phases. Also, we assume a normalized block time in the following, for example, $T = 1$, for convenience but without loss of generality.

6.3.1 Signal Model

Downlink WET. At each channel use of the downlink WET phase, the HAP transmits $K \leq \min(N, M)$ independent energy symbols $\{x_k\}$ such that the RF signal received by S_i (neglecting noise) is given by

$$y_i = \sqrt{\varrho_i} \mathbf{h}_i^T \sum_{k=1}^{K} \mathbf{w}_k x_k, \tag{6.21}$$

where $\mathbf{w}_k \in \mathbb{C}^{M \times 1}$ represents the precoding vector associated to x_k, thus $\sum_{k=1}^{K} ||\mathbf{w}_k||^2$ represents the total transmit power. Then, the energy harvested by S_i is given by

$$E_i^h = (1 - \tau_0) g\left(\varrho_i \left| \mathbf{h}_i^T \sum_{k=1}^{K} \mathbf{w}_k \right|^2 \right), \tag{6.22}$$

where g is the EH transfer function (see Section 2.2), assumed the same for all the IoT devices as in case they are equipped with the same EH circuitry.

Uplink WIT. Upon request, only N' (out of N) EH devices are scheduled for transmission in the uplink WIT phase. Let us assume p_i is the transmit power of S_i, and that all the active devices send a data message comprising b information bits per Hz, either using TDMA or spatial-division multiple access (SDMA) schemes as described next.

- TDMA: In the TDMA scheduling, each active (scheduled for transmission) node S_i is assigned an equal portion $\tau_i = \tau_0/N'$ of the whole block without time-sharing. Since each device operates alone, an MRC filter, for instance, $\mathbf{h}_i^*/||\mathbf{h}_i||$, is optimal and thus used at the HAP. Therefore, the data message transmitted by S_i is received at the HAP with SNR given by

$$\gamma_i = \frac{p_i \varrho_i}{\Upsilon\sigma^2}\left|\frac{\mathbf{h}_i^H\mathbf{h}_i}{||\mathbf{h}_i||}\right|^2 = \frac{p_i\varrho_i}{\Upsilon\sigma^2}||\mathbf{h}_i||^2 = p_i\bar{\gamma}_i^{\text{tdma}}, \qquad (6.23)$$

where σ^2 and Υ are defined as in Section 6.2, and $\bar{\gamma}_i^{\text{tdma}} = \varrho_i||\mathbf{h}_i||^2/(\Upsilon\sigma^2)$. Note that power, but not time, allocation will be optimized[3].

- SDMA: In the SDMA setup, all active users are simultaneously served by leveraging the multiple antennas at the HAP, thus $\tau_i = \tau_0, \forall i$ such that S_i is active. We assume that $N' \leq M$, otherwise a more evolved implementation using SIC or hybrid TDMA-SDMA would be required, which we leave for future work. To decouple the simultaneous transmission, we use a zero forcing (ZF) combiner, which is simpler and more tractable than minimum mean square error (MMSE), thus the resulting equivalent SNR is given by [240, Eq.(26)]

$$\gamma_i = \frac{p_i\varrho_i}{\Upsilon\sigma^2}z_i = p_i\bar{\gamma}_i^{\text{sdma}}, \qquad (6.24)$$

where $\bar{\gamma}_i^{\text{sdma}} = \varrho_i z_i/(\Upsilon\sigma^2)$, and $z_i = 1/Z_{i,i}$ with

$$\mathbf{Z} = \left([\mathbf{h}_1, \mathbf{h}_2, \cdots, \mathbf{h}_{N'}]^H[\mathbf{h}_1, \mathbf{h}_2, \cdots, \mathbf{h}_{N'}]\right)^{-1}. \qquad (6.25)$$

6.3.2 Problem Formulation

Consider that the circuit energy consumption per block is fixed and equal to E_c. Then, each empty-buffer device requires a harvested energy that satisfies $E_i^h \geq E_c$, while an active device requires an extra $p_i\tau_i$ energy units to be able to transmit the data in the uplink WIT phase. Aiming to minimize the HAP's energy consumption, while guaranteeing the EH and data transmission requirements of the IoT devices, we can state the following optimization problem

---

3 The adoption of a power instead of a time optimization framework seems more practical as synchronization issues are alleviated. However, we recognize that an additional per-user time allocation optimization in TDMA constitutes an interesting and worth addressing research direction for the scenario considered here.

$$\textbf{P8:} \quad \min_{\substack{\mathbf{w}_k \in \mathbb{C}^{M \times 1}, \\ \{p_i\}, \tau_0}} \quad f(\mathbf{w}_k, p_i, \tau_0) = (1 - \tau_0) \sum_{k=1}^{N'} ||\mathbf{w}_k||^2 \tag{6.26a}$$

$$\text{s. t.} \quad E_i^h \geq E_c, \quad \forall i \in \{i | S_i \text{ is inactive}\}, \tag{6.26b}$$

$$E_i^h \geq E_c + p_i \tau_i, \quad \forall i \in \{i | S_i \text{ is active}\}, \tag{6.26c}$$

$$\tau_i \log_2(1 + \bar{\gamma}_i p_i) \geq b, \quad \forall i \in \{i | S_i \text{ is active}\}, \tag{6.26d}$$

$$\begin{cases} \tau_i = \tau_0/N', & \text{if TDMA} \\ \tau_i = \tau_0, & \text{if SDMA} \end{cases}, \tag{6.26e}$$

$$0 \leq \tau_0 < 1, \tag{6.26f}$$

where $\bar{\gamma}_i \in \{\bar{\gamma}_i^{\text{tdma}}, \bar{\gamma}_i^{\text{sdma}}\}$ according to the adopted scheduling scheme. It can be observed by simple inspection of (6.26a), (6.26b), (6.26c) and (6.26d) that **P8** is not convex in its current form.

Note that the data rate constraint in (6.26d) relies on Shannon capacity formulation, which does not hold tight for short-message transmissions, but in turn eases the analysis. A more accurate formulation considering short packet transmissions would result from using the finite blocklength formulation (A.4)[4]. However, the complexity of the corresponding optimization would significantly scale, thus we avoid such an approach here, although we note that it constitutes an interesting future research direction.

6.3.3 Optimization Framework

Since in practice time granularity is limited, for example, by clock and synchronization issues of the hardware, we may have a limited set $\mathcal{T}$ of possible values of τ_0 (and consequently of τ_i). Thus, we adopt a simple, yet practical, optimization approach based on minimizing $f(\mathbf{w}_k, p_i, \tau_0)$ for each fixed $\tau_0 \in \mathcal{T}$, after which we select the τ_0 that leads to the general optimum solution. This is

$$\tau_0^{\text{opt}} = \arg \min_{\tau_0 \in \mathcal{T}} \min_{\{\mathbf{w}_k\}, \{p_i\}} \{f(\mathbf{w}_k, p_i, \tau_0) \mid \tau_0\}, \tag{6.27}$$

$$\{\mathbf{w}_k^{\text{opt}}\}, \{p_i^{\text{opt}}\} = \arg \min_{\{\mathbf{w}_k\}, \{p_i\}} f(\mathbf{w}_k, p_i, \tau_0^{\text{opt}}). \tag{6.28}$$

---

4 For instance, by finding the SNR threshold γ_{th} that guarantees a desired decoding error probability ε_0 when transmitting the data message over $n = \lfloor \tau_i T/T_c \rfloor$ channel uses, for instance, by using Algorithm 3 in Section 6.2.2, and setting $\bar{\gamma}_i p_i \geq \gamma_{\text{th}}$.

The time-fixed optimization can be simplified by noticing that

$$p_i^{\text{opt}} \mid \tau_i = \frac{2^{b/\tau_i} - 1}{\bar{\gamma}_i}. \tag{6.29}$$

Now, let us rewrite E_i^h in (6.22) as

$$E_i^h = (1 - \tau_0) g\Big(\varrho_i \operatorname{Tr}(\mathbf{W} \mathbf{H}_i) \Big), \tag{6.30}$$

as was similarly done in Section 4.2. Herein, $\mathbf{W} = \sum_{k=1}^{K} \mathbf{w}_k \mathbf{w}_k^H$ and $\mathbf{H}_i = \mathbf{h}_i \mathbf{h}_i^H$. Notice that $\mathbf{W}$ is a Hermitian matrix with maximum rank $\min(N, M)$. Then, $\mathbf{W}$ can be found for fixed-time and fixed-power (according to (6.29)) by solving the following SDP:

$$\textbf{P9:} \quad \min_{\mathbf{W} \in \mathbb{C}^{M \times M}} \quad \operatorname{Tr}(\mathbf{W}) \tag{6.31a}$$

$$\text{s.t.} \quad \operatorname{Tr}(\mathbf{W} \mathbf{H}_i) \geq \frac{1}{\varrho_i} g^{-1}\left(\frac{E_c + a_i p_i \tau_i}{1 - \tau_0} \right), \ \forall i, \tag{6.31b}$$

$$\mathbf{W} \succeq 0, \tag{6.31c}$$

where $a_i = 1$ if S_i is active, and $a_i = 0$ otherwise. Note that IPMs are usually adopted to efficiently solve SDP problems [220, 221]. In this case, it can be shown that solving **P9** requires around $\mathcal{O}(\sqrt{M} \ln(1/\varepsilon))$ iterations, with each iteration requiring at most $\mathcal{O}(M^6 + NM^2)$ arithmetic operations [221], and where ε represents the solution accuracy attained when the algorithm ends.

The proposed network energy minimization (NEM) framework is summarized in Algorithm 4. Based on the complexity analysis of **P8**, we can conclude that Algorithm 4 requires $\mathcal{O}(|\mathcal{T}| \sqrt{M} \ln(1/\varepsilon))$ iterations with $\mathcal{O}(|\mathcal{T}|(M^6 + NM^2))$ arithmetic operations per iteration since the optimization scales linearly with $|\mathcal{T}|$.

6.3.4 TDMA versus SDMA

Next, we present numerical results on the performance of the TDMA and SDMA scheduling. Unless stated otherwise, results are obtained by setting $M = 8$, $\sigma^2 = -90$ dBm, $b = 1$ bps/Hz, $E_c = 10$ μJ, $\varrho = 54$ dB for an entire set of $N = 64$ EH devices, which may correspond to the path loss experienced at a distance of 25 m when using the log-distance model in (4.21) and (5.2), and

---
Algorithm 4 NEM for multi-user WPCN

1: **Input:** $\{h_i\}$, $\{\varrho_i\}$, $\{\eta_i\}$, E_c, b, σ^2 and $\mathcal{T}$
2: **for** $\tau_0 \in \mathcal{T}$ **do**
3:     Compute $p_i(\tau_0)$ according to (6.29)
4:     Solve **P9** $\rightarrow$ Output: $\mathbf{W}(\tau_0)$
5: **end for**
6: Set $\tau_0^{\text{opt}} = \arg\min_{\tau_0}(1 - \tau_0)\mathbf{W}(\tau_0)$
7: Set $p_i^{\text{opt}} = p_i(\tau_0^{\text{opt}})$, $\mathbf{W}^{\text{opt}} = \mathbf{W}(\tau_0^{\text{opt}})$
8: Set $\{\mathbf{w}_k^{\text{opt}}\}$ as the eigenvectors of $\mathbf{W}^{\text{opt}}$
9: **Output:** $\{\mathbf{w}_k^{\text{opt}}\}$, τ_0^{opt}, $\{p_i^{\text{opt}}\}$

$\Upsilon = 9.8$ dB (as in Section 6.2.3). Moreover, we assume the devices are active with a given activation probability, which unless stated otherwise is set to 0.1. However, for those scenario instances where the number of active devices is greater than the number of antennas M, the set of active devices is randomly reduced such that its cardinality matches M. Finally, we adopt a Rician quasi-static fading model with LOS factor of 10 dB, and the sigmoidal (non-linear) EH model in (4.25) [100] with $\varpi_2 = 3 \times 10^{-4}$, $c_0 = 2 \times 10^{-4}$, and $c_1 = 6000$, which produces the RF–DC energy transfer and efficiency curves illustrated in Figure 6.7.

On the Impact of the Devices' Activity. Figure 6.8 illustrates the system performance as a function of the devices' activation probability. As such probability increases, more devices tend to contend for the spectrum resources, thus the HAP's transmit power and WET time increase to satisfy their energy demands. Observe that as the activation probability becomes smaller, the system performance under SDMA and TDMA converges. This is because the number of instantaneously active devices decreases to 1 or even 0, for which there is no difference in the operation of both schemes. However, as the activation probability increases, SDMA becomes more suitable as it exploits more efficiently the available spatial resources to serve the potentially relatively large (with respect to the number of HAP antennas) number of active devices.

On the Impact of the Number of Transmit Antennas. Figure 6.9 demonstrates that SDMA loses its appeal relative to TDMA as M decreases. Remember that the adopted TDMA design does not optimally allocate the user-specific time resources, thus, by incorporating such an extra degree of freedom, a greater performance gap with respect to SDMA is expected, especially for small M.

On the Impact of the Circuit Power Consumption. As expected, the greater the circuit power consumption of the EH devices E_c is, the greater their energy

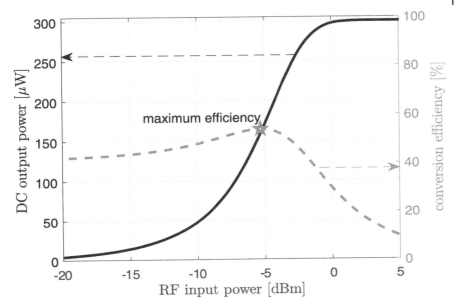

Figure 6.7 EH transfer function and conversion efficiency for $\varpi_2 = 3 \times 10^{-4}$, $c_0 = 2 \times 10^{-4}$, and $c_1 = 6000$.

demands are, and consequently the greater the PB's energy consumption become to efficiently serve them. Interestingly, the performance gap between both TDMA and SDMA slowly decreases as E_c increases. This is because the circuit energy become comparable to the transmit energy requirements, and even starts to dominate the devices' energy consumption, thus the WIT scheduling loses performance impact.

6.4 Cognitive MAC

The physical layer performance of TDMA and SDMA scheduling schemes was evaluated in the previous section. Such basic MAC mechanisms may suffice in the case of operating in unlicensed spectrum bands and as long as the HAP is powered by regular (wired) energy sources. However, when such conditions do not hold, for instance, when the HAP is powered by renewable energy sources and spectrum rights belong to another node, for example, a primary user (PU), more evolved MAC strategies are required for optimum performance under interference and energy availability constraints. In such scenarios, the time for the HAP to send an RF charging signal may be opportunistic since the arrival

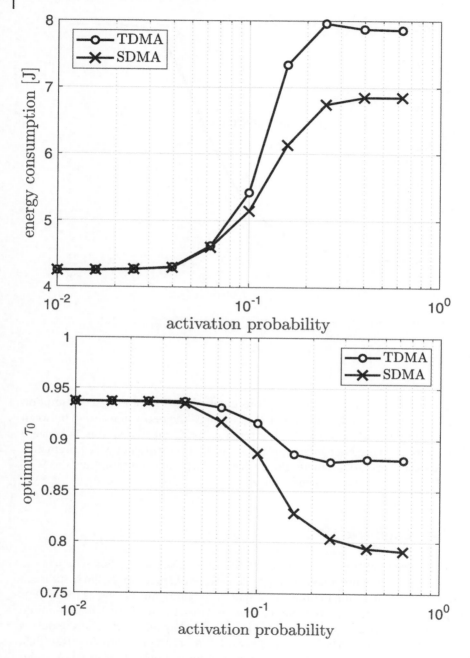

Figure 6.8 (a) Optimum energy consumption (top) and (b) associated WIT duration (bottom) as a function of the activation probability.

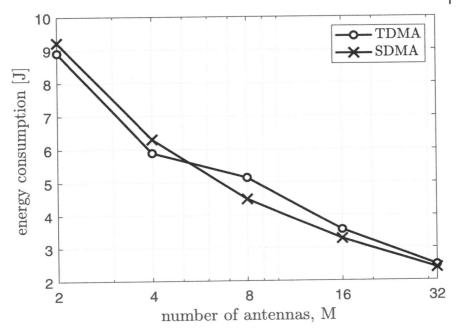

Figure 6.9 Optimum energy consumption as a function of the number of transmit antennas.

time of the PU may be unpredictable. Therefore, cognitive radio (CR) technology may be needed, especially if no specific coordination between the PU and HAP exists[5].

In this section, we comment on suitable data link layer mechanisms for a green WPCN operating in the coverage area of a PU service. Our discussions are mainly based on the recent contribution from Tzeng *et al.* in [241]. The CR WPCN considered here for illustration purposes is shown in Figure 6.11. The HAP, powered by green energy sources, performs spectrum sensing operations and broadcasts the results to its associated set $\mathcal{S} = \{S_i\}$ of EH devices through a dedicated control channel. Since the EH devices are assumed to be in the neighborhood of the HAP, they do not need to equip a module of spectrum sensing, thus reducing their complexity and costs. According to the sensing result received from the HAP, the EH devices would determine whether to enter the modes of data transmission (uplink), EH (downlink), or sleep. We focus on a

---

5 Note that if such a coordination is possible, the PU receiver can be configured/tuned to remove the interference contribution from quasi-(or fully-)deterministic WET signals as discussed in Section 2.4.6.

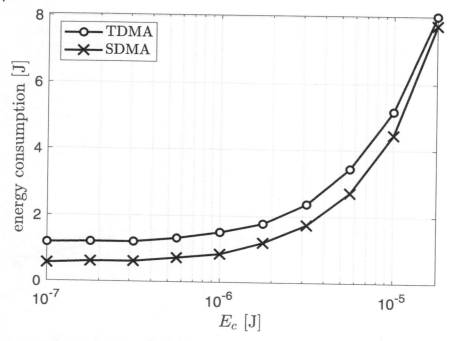

Figure 6.10 Optimum energy consumption as a function of the devices fixed energy consumption.

single frequency band operation, although all the procedures discussed next can be easily extrapolated to a more practical multi-band system.

If the sensing outcome indicates that the radio band is busy, a device S_i could enter a sleep mode to reduce energy consumption. Otherwise, S_i starts a MAC protocol before transmitting its data. In addition, the HAP is equipped with an FD transmission module, which allows spectrum sensing operations and message broadcasting to be performed simultaneously.

6.4.1 Time Sharing and Scheduling

Figure 6.12 illustrates the adopted time structure consisting of time slots, frames, and multi-frames. The HAP senses the spectrum at the beginning of a multi-frame. If the sensing outcome is idle, a MAC protocol is enabled in this multi-frame; otherwise, the multi-frame is marked as occupied and cannot be used[6]. In the case when an idle state has been sensed, the HAP and EH

---

6 This is the overlay CR paradigm, where the secondary users use the licensed spectrum only if it is not being already used by the PUs. Alternatively, users may adopt the underlay CR protocol where concurrent transmissions are allowed but under strict transmit power constraints to

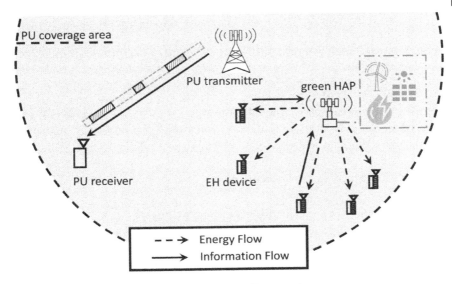

Figure 6.11 Green wireless powered cognitive radio network.

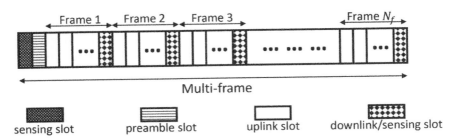

Figure 6.12 Illustration of the time sharing and scheduling structure.

devices use the preamble time slot for synchronization. Next, N_f frames follow the preamble time slot, where each frame includes N_d uplink time slots and each uplink slot carries one uplink data packet. An EH device S_i may send a packet in an uplink slot j, $j = 1, 2, \cdots, N_d$, in a certain frame, and then, the slots at the same position j in successive frames are reserved for such a device until the current multi-frame ends. Meanwhile, one downlink slot is reserved to broadcast information from the HAP to the EH devices at the end of each frame. Note that the maximum number of frames in a multi-frame is limited in order to prevent devices from occupying slots for a long time; that is,

---

avoid strong interference to the PU receivers. Note also that in a practical implementation where multiple frequency bands may be available, the HAP may also switch to a different radio band and repeat the sensing procedure.

the maximum number of frames shall consider the fairness between devices [241, 242].

Observe that PUs may return during the period of a multi-frame. Thus, at the downlink time slot, and exploiting the FD radio, the HAP senses again the spectrum to detect whether the PUs became active. If PU activity is not detected, the secondary network (composed of the HAP and the EH devices) continues its operation normally in the current multi-frame; otherwise, the current multi-frame ends. In addition to data transmission/reception tasks, the HAP may receive a charging request from a certain or several EH devices.

6.4.2 MAC Protocol at the Device Side

Assume that the EH devices already know whether the frame is idle or not. If the frame is not idle, there is nothing they can do but to wait for an idle state flag (reported by the HAP). If the frame is idle and a device S_i has already reserved a slot for uplink transmissions, it waits for it at each frame of the multi-frame, otherwise, if S_i requires sending a data message (either an uplink data message or a charging request) but does not have a reserved slot, it enters a contention state. The detailed specifications of the MAC protocol executed at S_i are exposed in Algorithm 5.

Note that the EH devices are aware of the slots that are already reserved as they have been announced by the HAP in the downlink slots. Then, in the contention state, a device S_i attempts to contend for an un-reserved slot, either for data transmission (if the residual energy B_s in the battery is above a predefined energy threshold ξ) or for a charging request (if the residual energy is too low to support future data transmission rounds and local processing tasks, but sufficiently high, above a sleep threshold φ_s, to support a single transmission). The eventual transmission occurs with a certain predefined probability p_c, and note that a sleep threshold is also used here to prevent transmissions exhausting low battery energy levels.

As S_i sends a packet in an un-reserved slot, it leaves the contention state and waits an acknowledgment (ACK) from the HAP in the downlink slot. Once a positive ACK has been received, all the slots at the same position in the successive frames are reserved for S_i, which is now in a reservation state. Otherwise, the device backs-off a random number k of slots and contends for a slot again once the back-off timer expires. After backing-off k slots, where k is chosen randomly in $[1, 2, \cdots, K]$ with [241, Eq.(1)]

$$K = \frac{1}{\alpha_1} \left[2^{\frac{\ln(n_c \phi / t_s)}{\alpha_2}} - 1 \right],$$

(6.32)

S_i contends for a slot again. In (6.32), n_c is the number of collisions after last successful transmission, ϕ is a QoS parameter (e.g., packet delay or delay jitter), t_s is the duration of the time slots, and α_1 and α_2 are scaling factors.

Algorithm 5 MAC protocol at each frame of an idle multi-frame – S_i side

---

Input: charge threshold ξ, sleep threshold φ_s, contention probability p_c, residual energy in the battery B_s, maximum number of back-off slots K

Initialize number of back-off slots: $k = 0$

Uplink slot $j = 1, \cdots, N_d$

1: **if** slot is reserved for S_i and $B_s \geq \xi$ **then** transmit the data message;
2: **else if** slot is reserved for S_i and $B_s \geq \varphi_s$ **then** transmit a charging request;
3: **else if** S_i is in contention, $B_s \geq \xi$ and $k = 0$ **then** transmit the data message with probability p_c;
4: **else if** S_i is in contention, $B_s \geq \varphi_s$ and $k = 0$ **then** transmit a charging request with probability p_c;
5: **else if** $k > 0$ **then** $k \leftarrow k - 1$ **end**
6: **if** no data message or charging request transmission **then** harvest RF energy **end**
7: **if** a broadcast message from HAP is corrupted **then** interrupt activities in the current multi-frame;
8: **else if** positive ACK is received for a message sent in the current slot **then** reserve such a slot in successive frames of current multi-frame;
9: **else if** negative ACK (or not ACK at all) **then** set $k \leftarrow$ random integer number in $[1, K]$ **end**

Downlink slot

1: Harvest RF energy and decode signaling information data using a corresponding receive SWIPT mechanism

---

Finally, note that the EH devices expect to receive broadcast messages (data and/or signaling information as specified above) from the HAP at downlink slots, at the same time that some of them may require RF charging services. Therefore, the HAP operates with a SWIPT mechanism and the EH devices implement the corresponding joint receive information and energy protocol. These issues are discussed later in Chapter 7. A few other practical considerations that are not illustrated in Algorithm 5 are for instance:

- data messages exceeding their associated deadline are removed from data buffer queues;
- when devices receive corrupted (undecodable) downlink messages from the HAP, they interrupt their transmission operations in the current multi-frame.

6.4.3 MAC Protocol at the HAP Side

In an idle multi-frame, the HAP can perform several tasks such as uplink data reception, downlink WET and signaling broadcast, and spectrum sensing to address an eventual return of the PU service. The associated procedures are discussed next and illustrated in Algorithm 6.

Data Reception. For a jth slot in a frame, if the HAP successfully decodes a data message from a device S_i, all the jth slots in successive frames are reserved for S_i. These allocations are announced in downlink slots. The HAP releases a reserved slot when a data message was not received therein, which happens when the device did not transmit because there was no longer any data to transmit or because the residual energy in its battery was less than or equal to the sleep threshold.

WET Phase. The HAP keeps track of all the charging requests received during the uplink slots transmissions. It might be possible that the energy needs of all the low-battery devices cannot be met in a single downlink slot, especially given

- green energy availability issues at the HAP that limit its transmit power; and
- EH circuit limitations, e.g., saturation phenomenon, that prevents the devices from harvesting large amounts of RF power to compensate for a relative short WET phase.

In such a case, the HAP may continue serving such devices with uncompleted charging processes in successive frames and according to its energy availability specified by B_hap. A simple strategy is to cease WET services when B_hap becomes limited, which could be specified by a HAP's battery level threshold φ_hap. Note that several devices can simultaneously benefit from a common charging process, especially by taking advantage of EB from a multi-antenna HAP. The specificities of such a physical layer design are out of the scope of our discussions in this section, but in general can be implemented using diverse multi-antenna mechanisms such as those discussed in Chapter 4.

Handling PU Returns. It is possible that the radio band is idle at the beginning of a multi-frame but the PU service returns at some point before the multi-frame ends. To avoid the PU to be interfered by either the RF charging and signaling signals transmitted by the HAP in downlink slots or the data transmission of devices in uplink slots, the HAP takes advantage of the FD module to monitor the spectrum at downlink slots. If the PU services is detected, the HAP stops the transmissions.

6.5 Final Remarks

In this chapter, we studied WPCN scenarios. Specifically, we discussed the main WPCN models and studied the performance of 1. a single-antenna

---

Algorithm 6 MAC protocol at each frame of an idle multi-frame – HAP side

---

Input: set of reserved slots, HAP's battery level threshold φ_{hap}, residual energy in the battery of the HAP B_{hap}

Uplink slot $j = 1, \cdots, N_d$

1: **if** a data message is successively received from S_i **then** reserve the slot for S_i in the successive frames;

2: **else** release the slot in the successive frames **end**

Downlink slot

1: Sense the spectrum during the entire downlink slot

2: **if** PU service is detected **then** terminate the multi-frame;

3: **else if** $B_{\text{hap}} \geq \varphi_{\text{hap}}$ and there are pending-to-serve and/or new charging requests **then** implement joint WET and signaling transmissions (e.g., ACKs);

4: **else** signaling transmissions (e.g., ACKs) **end**

---

point-to-point system in terms of error probability and device power consumption, and 2. a multi-user network with multi-antenna capabilities under time-division or space-division multiplexing WIT protocols in terms of network energy consumption. Furthermore, we provided additional discussions on potential MAC mechanisms at the HAP and device side to deal with scenarios where the spectrum rights belong to another network service provider and the HAP is powered by unstable energy sources, for example, green energy sources.

Regarding the single-user system, we evaluated the system reliability and average transmit power of devices in scenarios where energy accumulation between transmission rounds were or were not allowed. For the former scenarios, we illustrated the performance of a power control protocol FEIPC using CSI at the IoT device side. Meanwhile, the HTT protocol, which exhausts all the energy in a device's battery for transmission at each round, was adopted for the latter scenarios and benchmarking. Results showed that a small, but practical, battery capacity may suffice for appropriate system performance, and that the FEIPC protocol considerably outperforms HTT both in terms of reliability and a device's power consumption. We illustrated that an efficient FEIPC design requires properly setting a target error probability threshold. In fact, we showed there is an optimum value of target error probability that minimizes the achievable error probability. However, the higher the target error probability is, the lower the energy consumption becomes.

Regarding the multi-user multi-antenna system, we described the functioning of a TDMA and SDMA WIT scheduling, and formulated a HAP energy minimization problem subject to users' rate constraints for both schemes. The optimization framework relies on a discretization of the time space for WET and WIT allocation, a simple, but optimal, power control strategy, and an SDP convex formulation to obtain the EB vectors at the HAP. Results evidenced that the system performance worsens as the device's activation probability and/or device's circuit power consumption increases, which may be counteracted by equipping the HAP with more antennas. For a small number of antennas at the HAP, TDMA was shown to excel, otherwise SDMA resulted to be the most appropriate scheduling approach.

7

Simultaneous Wireless Information and Power Transfer

7.1 SWIPT Schemes

With respect to SWIPT, three types of receiver schemes have been proposed and analyzed in recent years [131] (see Figure 7.1):

- A separate receiver is shown in Figure 7.1(a). The antenna array is divided into two sets, each connected to the EH circuitry or the information receiver. This scheme allows EH and ID to be performed independently and concurrently. In this case, the EH and ID circuits may belong to the same or different devices.
- A co-located receiver scheme permits the EH and information receiver to share the same antenna(s). This scheme can be split into two categories: 1. time switching (TS), and 2. power splitting (PS), which are shown in Figure 7.1(b) and Figure 7.1(c), respectively. When operating with TS, the network node switches and uses either the information receiver or the RF-EH circuit for the received RF signal; while when operating with PS, the received RF signal is split into two streams with different power levels, one for the information receiver and the other for the RF energy harvester.
- An integrated receiver scheme is shown in Figure 7.1(d). The implementation of RF-to-baseband conversion for ID is integrated with the EH circuit via the rectifier. The RF flow controller can also adopt a switcher or power splitter, like in the co-located receiver scheme.

Another SWIPT receiver not specifically illustrated in Figure 7.1 employs the so-called *spatial switching* (SS) technique discussed in [243]. The SS technique can be applied in MIMO configurations and achieves SWIPT in the spatial domain by exploiting the multiple degrees of freedom of the interference channel. Based on the eigen decomposition of the MIMO channel, the communication link is transformed into parallel eigenchannels that convey either information or energy.

Wireless RF Energy Transfer in the Massive IoT Era: Towards Sustainable Zero-energy Networks.
First Edition. Onel Alcaraz López and Hirley Alves.
© 2022 by The Institute of Electrical and Electronics Engineers, Inc. Published 2022 by John Wiley & Sons, Inc. Companion website: www.wiley.com/go/Alves/WirelessEnergyTransfer

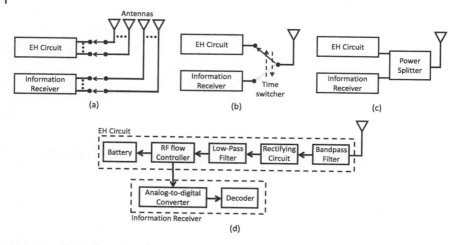

Figure 7.1 Receiver schemes for SWIPT: (a) separate receiver; (b) time switching; (c) power splitting; and (d) integrated receiver. *Source:* Adaptation from [157, Fig. 1].

7.2 Separate EH and ID Receivers

Let us consider a multi-user MISO SWIPT system with separate EH and ID receivers operating over a single frequency band. Specifically, a HAP, which is equipped with M antennas, serves a set $\mathcal{S}_{\text{id}}$ and $\mathcal{S}_{\text{eh}}$ of single-antenna ID and EH receivers (with $\mathcal{S}_{\text{id}} \cap \mathcal{S}_{\text{eh}} = \emptyset$), respectively in the downlink, such that $M \geq |\mathcal{S}_{\text{id}}|$. Without loss of generality, assume the HAP assigns one dedicated information/energy transmission beam to each EH/ID receiver. Therefore, the signal transmitted from the HAP is given by

$$\mathbf{x} = \sum_{i \in \mathcal{S}_{\text{id}}} \mathbf{v}_i s_i^{\text{id}} + \sum_{j \in \mathcal{S}_{\text{eh}}} \mathbf{w}_j s_j^{\text{eh}}, \tag{7.1}$$

where $\mathbf{v}_i, \mathbf{w}_j \in \mathbb{C}^{M \times 1}$ is the precoding vector for the ith ID and the jth EH receiver, respectively; similarly s_i^{id} and s_j^{eh} is the information-bearing signal and energy-carrying signal, respectively. For information signals, Gaussian inputs are assumed, thus the s_i^{id} are i.i.d. zero-mean unit-variance CSCG random variables. Meanwhile, for energy signals, it can be any arbitrary signal since no information is carried. However, in practice, the signal waveform design must comply with certain power spectral density regulations on microwave radiation, thus choosing an appropriate distribution is also relevant because of the waveform-dependent non-linear EH impact, which will be discussed later in Section 7.4.1. Herein, we just assume that the s_j^{eh} are independent white sequences from an arbitrary distribution with $\mathbb{E}\left[|s_i^{\text{eh}}|^2\right] = 1$.

We assume a quasi-static fading environment and denote $\mathbf{h}_i^{\mathrm{id}}, \mathbf{h}_j^{\mathrm{eh}} \in \mathbb{C}^{M \times 1}$ as the channel vector from the HAP to the ith ID and jth EH device, respectively. Moreover, channel realizations corresponding to different users are assumed to be independently distributed, and perfect CSI is available at the HAP and ID devices.

Finally, we would like to note that the scenario and discussions presented in this section are mainly based on those exposed in the work of Xu *et al.* [244], although we deviate from this pioneering work in several aspects, for instance, we assume a generic EH model, imperfect cancellation of deterministic WET signals at the ID devices, and more importantly, we aim to minimize the HAP power consumption subject to QoS requirements of the IoT devices instead of maximizing the total energy harvested by the IoT network subject to a specific HAP transmit power budget and ID QoS constraints as in [244].

7.2.1 Problem Formulation

The discrete-time baseband signal at the ith ID receiver is given by

$$y_i^{\mathrm{id}} = \mathbf{x}^H \mathbf{h}_i^{\mathrm{id}} + u_i, \tag{7.2}$$

where $u_i \sim \mathcal{CN}(0, \sigma_i^2)$ is the i.i.d. AWGN. Meanwhile, the corresponding SINR can be expressed as

$$\gamma_i = \frac{|\mathbf{v}_i^H \mathbf{h}_i^{\mathrm{id}}|^2}{\sum_{\forall k \neq i} |\mathbf{v}_k^H \mathbf{h}_i^{\mathrm{id}}|^2 + \delta_i \sum_{\forall j} |\mathbf{w}_j^H \mathbf{h}_i^{\mathrm{id}}|^2 + \sigma_i^2}, \tag{7.3}$$

where $\delta_i \in [0, 1]$ denotes the SIC coefficient of the WET signal. Note that since energy beams carry no information but instead pseudorandom signals whose waveforms can be assumed to be known at both the HAP and each ID receiver prior to data transmission, their resulting interference can be canceled if this additional operation is implemented. Therefore, herein $\delta_i > 0$ means that the cancellation operation is implemented at the ith ID device, which may be assumed to be perfect by setting $\delta_i = 1$, $\forall i$; while one can easily emulate the scenario where such cancellation functionality is not implemented by letting $\delta_j = 0$.

At the jth EH device, the harvested power is given by

$$E_j^h = g\left(\sum_{\forall i} |\mathbf{v}_i^H \mathbf{h}_j^{\mathrm{eh}}|^2 + \sum_{\forall l} |\mathbf{w}_l^H \mathbf{h}_j^{\mathrm{eh}}|^2 \right), \tag{7.4}$$

where g is the EH transfer function (see Section 2.2), assumed the same for all the IoT devices as if they were equipped with the same EH circuitry.

Herein, we focus on minimizing the HAP transmit power subject to individual EH requirements specified by $\{\xi_j\}$, where each ξ_j belongs to the image of g (i.e., EH requirements are reachable with the provided EH circuit), and SINR constraints given by $\{\gamma_i^o\}$, as illustrated in **P10**.

$$\textbf{P10:} \quad \min_{\mathbf{v}_i, \mathbf{w}_j \in \mathbb{C}^{M \times 1}, \forall i,j} \sum_{\forall i} ||\mathbf{v}_i||^2 + \sum_{\forall j} ||\mathbf{w}_j||^2 \tag{7.5a}$$

$$\text{s. t.} \quad E_j^h \geq \xi_j, \quad \forall j, \tag{7.5b}$$

$$\gamma_i \geq \gamma_i^o, \quad \forall i. \tag{7.5c}$$

Note that although the optimization function is quadratic, thus, convex, the set of $|\mathcal{S}_{\text{id}}| + |\mathcal{S}_{\text{eh}}|$ constraints specified by (7.5b) and (7.5c) are not, hence, **P10** is non-convex. Independently of this, we can still make the following claim

Lemma 7.1 **P10** is feasible if and only if

$$\sum_{i \in \mathcal{S}_{\text{id}}} \frac{\gamma_i^o}{1 + \gamma_i^o} \leq \text{rank}(\bar{\mathbf{H}}^{\text{id}}), \tag{7.6}$$

where $\bar{\mathbf{H}}^{\text{id}} = [\mathbf{h}_1^{\text{id}}, \mathbf{h}_2^{\text{id}}, \cdots, \mathbf{h}_{|\mathcal{S}_{\text{id}}|}^{\text{id}}]$.

Proof. Without SINR requirements, **P10** is always feasible as the WET signal's power can be deliberately increased until (7.5b) is satisfied. Let us assume now the presence of only ID devices with their corresponding SINR requirements, thus the HAP does not transmit dedicated WET signals, i.e., $\mathbf{w}_j = \mathbf{0}$, $\forall j$. Then, the resulting problem is composed by the first addend of (7.5a) and constraint (7.5c), which is known to be feasible only when (7.6) is satisfied [245, Theorem III.1]. The incorporation of EH devices to the system, and their corresponding energy requirements does not impact (7.6). Finally, incorporating dedicated WET signals constitutes just another degree of freedom that may be exploited to further improve the system performance by allowing (7.5b) to be satisfied more easily. ∎

Hereinafter, we assume a set of SINR constraints and channel realizations that satisfy (7.6), hence feasible optimization problems.

7.2.2 Optimal Solution

Since the constraints (7.5b) and (7.5c) can be easily put in a quadratic form, we can conclude that **P10** is a non-convex quadratically constrained quadratic program (QCQP), for which semi-definite relaxation (SDR) optimization techniques perform efficiently [246]. In order to proceed further, let us define the following matrices $\mathbf{V}_i = \mathbf{v}_i \mathbf{v}_i^H$ and $\mathbf{W} = \sum_{\forall j} \mathbf{w}_j \mathbf{w}_j^H$, thus, with maximum ranks 1 and $\min\left(M, |\mathcal{S}_{\text{eh}}|\right)$, respectively. By dropping such rank constraints, the SDR of **P10** is given by

$$
\mathbf{P11:} \quad \min_{\mathbf{W}, \{\mathbf{V}_i\} \in \mathbb{C}^{M \times M}} \sum_{\forall i} \text{Tr}(\mathbf{V}_i) + \text{Tr}(\mathbf{W}) \tag{7.7a}
$$

$$
\text{s.t.} \quad \sum_{\forall i} \text{Tr}(\mathbf{H}_j^{\text{eh}} \mathbf{V}_i) + \text{Tr}(\mathbf{H}_j^{\text{eh}} \mathbf{W}) \geq g^{-1}(\xi_j), \ \forall j, \tag{7.7b}
$$

$$
\frac{1}{\gamma_i^\circ} \text{Tr}(\mathbf{H}_i^{\text{id}} \mathbf{V}_i) - \sum_{\forall k \neq i} \text{Tr}(\mathbf{H}_i^{\text{id}} \mathbf{V}_k)
$$

$$
- \delta_i \text{Tr}(\mathbf{H}_i^{\text{id}} \mathbf{W}) \geq \sigma_i^2, \ \forall i, \tag{7.7c}
$$

$$
\mathbf{W}, \mathbf{V}_1, \cdots, \mathbf{V}_{|\mathcal{S}_{\text{id}}|} \succeq 0, \tag{7.7d}
$$

where $\mathbf{H}_i^{\text{id}} = \mathbf{h}_i^{\text{id}}(\mathbf{h}_i^{\text{id}})^H$ and $\mathbf{H}_j^{\text{eh}} = \mathbf{h}_j^{\text{eh}}(\mathbf{h}_j^{\text{eh}})^H$.

In fact, and as numerically illustrated in the next section, the optimal solution of **P11** (given by $\mathbf{W}^{\text{opt}}$ and $\{\mathbf{V}_i^{\text{opt}}\}$) meets the required rank requirements very often. In such cases, the optimal HAP precoders correspond to the eigenvectors of $\mathbf{W}^{\text{opt}}$ and $\{\mathbf{V}_i^{\text{opt}}\}$; otherwise one may choose only the strongest eigenvector(s) and project the solution into the optimization feasibility set.

7.2.3 Performance Results

Herein, we provide numerical examples on the performance of the designed SWIPT precoding strategy. We assume that the signal attenuation from the HAP to all the EH IoT devices is 54 dB, which may correspond to the pathloss experienced at a distance of 25 m when using the log-distance model in (4.21) and (5.2), and that to all ID devices is 70 dB (at $\sim$100 m). We adopt the EH transfer function in (4.25) [100] and set its parameters similar to the setup in Section 6.3.4, producing the result in Figure 6.7. Moreover, the $\mathbf{h}^{\text{id}}$ and $\mathbf{h}^{\text{eh}}$ channel vectors are randomly generated from i.i.d. Rician fading with a LOS factor of 0 dB and 10 dB, respectively. Unless stated otherwise, we set $M = 8$, $|\mathcal{S}_{\text{eh}}| = |\mathcal{S}_{\text{id}}| = 6$, $\sigma_i^2 = -90$ dBm, $\gamma_i^\circ = 10$ dB, $\forall i$, $\xi_j = 100 \ \mu\text{W}$, $\forall j$.

On the Solution Optimality. Figure 7.2(a) illustrates the rank of matrices $\mathbf{W}^{\mathrm{opt}}$ and $\{\mathbf{V}_i^{\mathrm{opt}}\}$, obtained from solving **P11**, as a function of the SIC coefficient δ_j; while Figure 7.2(b) shows the minimum average transmit power requirements of the HAP. Observe that $\mathbf{W}^{\mathrm{opt}}$ always satisfies its associated rank constraint, that is rank smaller than $\min(M, |\mathcal{S}_{\mathrm{eh}}|)$, while $\mathbf{V}_i^{\mathrm{opt}}$ does not satisfy the rank 1 constraint for at least one i only when $\delta_j = 0$. Thus, the corresponding precoder design using such matrices, for instance, using the strongest eigenvector of each $\mathbf{V}_i$, will not lead to the global optimum performance only when the WET signal's cancellation is perfect. Although not shown here, this was also extensively corroborated via other numerical results. Interestingly, the specific value of δ_j does not impact the system performance when $\delta_j > 0$. Finally, observe that the number of WET signals is mainly influenced by the value of δ_j and the SINR thresholds.

On the Impact of $|\mathcal{S}_{\mathrm{eh}}|$, $|\mathcal{S}_{\mathrm{id}}|$, M, γ_i° and ξ_j. Figure 7.3 shows the minimum average transmit power requirement of the HAP as the number of EH and ID devices increases. Observe that the power requirements increase exponentially with the number of devices, but at a rate that decreases as the number of HAP antennas gets bigger. For instance, when $\mathcal{S}_{\mathrm{id}} = \mathcal{S}_{\mathrm{eh}} = 4$, the optimum transmit power of the HAP, when equipped with $M = 4$ antennas, is ~ 34 dB, while such value decreases to ~ 28.5 dB and ~ 27 dB when M increases to 8 and 12, respectively.

Meanwhile, the performance in terms of the SINR threshold for different EH requirements is illustrated in Figure 7.4. Note that the HAP needs to increase the transmit power as both ID and EH QoS requirements become more stringent. The transmit power scales proportional to the EH threshold, while slowly increases with the SINR threshold. Interestingly, the contribution of the WET signal power to the overall HAP's transmit power has a maximum peak around $\gamma_i^\circ = 0$ dB, beyond which it starts decreasing until becoming null in the high data rate regime, that is $\gamma_i^\circ > 25$ dB. This is because the WIT precoders are already sufficiently strong to satisfy not only the SINR requirements, but also the energy demands of the EH devices, which by design experience more favorable channel propagation conditions.

7.3 Co-located EH and ID Receivers

Different from the discussions in the previous section, herein the focus is on a co-located SWIPT system, where a HAP serves a set $\mathcal{S}$ of devices performing both EH and ID in a given frequency band. Each device is subject to both EH and rate constraints, specified by ξ_i and r_i. Similar to the notation in the previous section, herein M is the number of antennas at the HAP such that $M \geq |\mathcal{S}|$, $\mathbf{h}_i \in \mathbb{C}^{M \times 1}$ denotes the quasi-static channel vector from the HAP to the ith device. Moreover, $\mathbf{q}_i$, $\mathbf{v}_i$, $\mathbf{w}_i$ denotes the precoder associated with the

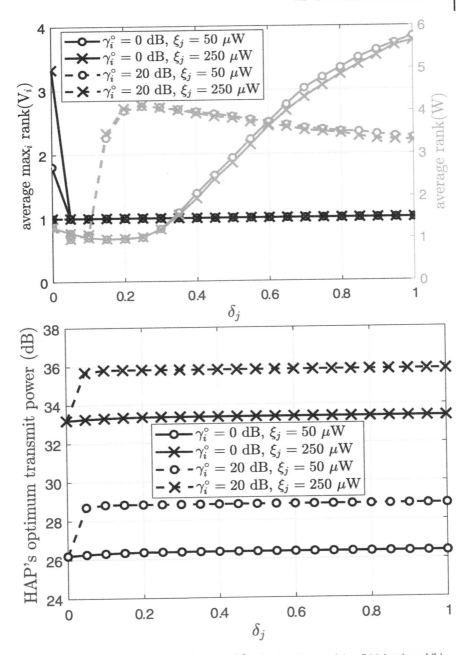

Figure 7.2 (a) Average rank of $\mathbf{W}^{\mathrm{opt}}$ and $\{\mathbf{V}_i^{\mathrm{opt}}\}$ obtained from solving **P11** (top), and (b) average optimum HAP transmit power (bottom) as a function of the SIC coefficient δ_j for $\gamma_i^\circ \in \{0,20\}$ dB and $\xi_j \in \{50,250\}$ μW.

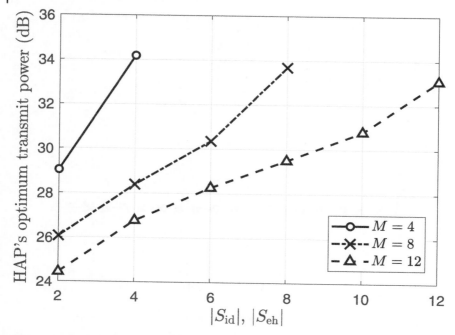

Figure 7.3 Average optimum HAP transmit power as a function of the number of EH and ID devices for $M \in \{4, 8, 12\}$.

unit-power ith SWIPT, dedicated WIT, and dedicated WET signal, s_i^{swipt}, s_i^{id}, s_i^{eh}, respectively. The latter two may or may not be needed depending on the specifically adopted SWIPT schemes discussed next.

7.3.1 Time Switching

Under the TS protocol, the HAP performs WET during a portion $\tau \in [0, 1]$ of a block time, and then executes WIT during the remaining, $1 - \tau$, portion of the block. Note that all the devices are necessarily synchronized with this timing.

Since WET and WIT occur disjointly, the energy and information signals received in the corresponding phases at the ith device are, respectively,

$$y_i^{\text{eh}} = \mathbf{h}_i^H \sum_{j \in \mathcal{S}} \mathbf{w}_j s_j^{\text{eh}}, \tag{7.8}$$

$$y_i^{\text{id}} = \mathbf{h}_i^H \sum_{i \in \mathcal{S}} \mathbf{v}_j s_j^{\text{id}} + u_{a,i} + u_{c,i}, \tag{7.9}$$

where $u_{a,i}$ is the AWGN due to the receiving antenna, and $u_{c,i}$ is the sampled AWGN due to RF band to baseband signal conversion. Their variances are

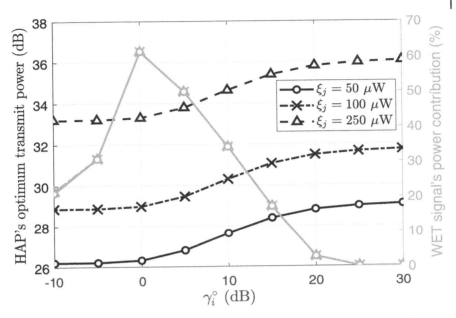

Figure 7.4 Average optimum HAP transmit power (left-hand y-axis) and corresponding power contribution of the WET signal (right-hand y-axis) as a function of the SINR threshold for $\xi \in \{50,100,250\}\ \mu$W.

denoted as $\sigma_{a,i}^2$ and $\sigma_{c,i}^2$, respectively. Observe that the effect of the noise in the EH signal is negligible and has been ignored.

Problem Formulation. Based on (7.8) and (7.9), the energy harvested by the ith EH device, and the SINR of the receive information signal, are respectively given by

$$E_i^h = \tau g\left(\sum_{j\in\mathcal{S}} |\mathbf{h}_i^H\mathbf{w}_j|^2\right), \tag{7.10}$$

$$\gamma_i = \frac{|\mathbf{h}_i^H\mathbf{v}_i|^2}{\sum_{k\in\mathcal{S}\backslash i} |\mathbf{h}_i^H\mathbf{v}_k|^2 + \sigma_{a,i}^2 + \sigma_{c,i}^2}, \tag{7.11}$$

where g is the EH transfer function (see Section 2.2), assumed to be the same for all the IoT devices as if they were equipped with the same EH circuitry. Then, the QoS constraints are given by

$$\text{EH: } E_i^h \geq \xi_i \rightarrow \qquad \sum_{j\in\mathcal{S}} |\mathbf{h}_i^H\mathbf{w}_j|^2 \geq g^{-1}(\xi_i/\tau) \triangleq \xi_i^\circ, \tag{7.12}$$

$$\text{ID: } (1-\tau)\log_2\left(1+\frac{\gamma_i}{\Upsilon}\right) \geq r_i \rightarrow \qquad \gamma_i \geq \left(2^{\frac{r_i}{1-\tau}}-1\right)\Upsilon \triangleq \gamma_i^\circ, \tag{7.13}$$

where Υ represents the SNR gap from the AWGN channel capacity due to a practical modulation and coding scheme being used.

Herein, we focus on minimizing the HAP transmit energy consumption, thus, the precoding optimization problem can be written as

$$
\textbf{P12:} \quad \min_{\mathbf{w}_i, \mathbf{v}_i \in \mathbb{C}^{M \times 1}, \forall i} \quad \tau \sum_{\forall i \in \mathcal{S}} ||\mathbf{w}_i||^2 + (1 - \tau) \sum_{\forall i \in \mathcal{S}} ||\mathbf{v}_i||^2 \tag{7.14a}
$$

$$
\text{s. t.} \quad \text{constraints (7.12), (7.13)}, \ \forall i \in \mathcal{S}. \tag{7.14b}
$$

Similar to **P10**, the feasibility of **P12** is delimited by (7.6). However, note that as τ decreases, ξ_i / τ increases and may more easily fall outside the domain of the image of g, thus making it impossible to meet the EH constraints.

Optimal Solution. Observe that **P12** can be split into two independent sub-problems dealing with the WET and WIT optimization separately since τ is assumed fixed. The WET sub-problem comprises the first addend of (7.14a), as the objective function, subject to the EH constraints in (7.12), thus, it can be efficiently solved by re-casting it as an SDP as

$$
\textbf{P13:} \quad \min_{\mathbf{W} \in \mathbb{C}^{M \times M}} \quad \text{Tr}(\mathbf{W}) \tag{7.15a}
$$

$$
\text{s. t.} \quad \text{Tr}(\mathbf{H}_i \mathbf{W}) \geq \xi_i^\circ, \qquad \forall i \in \mathcal{S}, \tag{7.15b}
$$

where $\mathbf{W} = \sum_{\forall j \in \mathcal{S}} \mathbf{w}_j \mathbf{w}_j^H$ is a Hermitian matrix with maximum rank given by $\max(M, |\mathcal{S}|)$, and $\mathbf{H}_i = \mathbf{h}_i \mathbf{h}_i^H$. Then, the optimum energy precoding vectors $\{\mathbf{w}_j^{\text{opt}}\}$ can be obtained as the eigenvectors of $\mathbf{W}$. Meanwhile, the WIT subproblem comprises the second addend of (7.14a), as the object function, subject to the SINR constraints in (7.13). This is a non-convex optimization problem, which fortunately can be put in a convex form by reformulating the SINR constraints as second-order cone constraints [247]. In fact, by denoting by $\mathbf{V} = [\mathbf{v}_1, \mathbf{v}_2, \cdots, \mathbf{v}_{|\mathcal{S}|}] \in \mathbb{C}^{M \times |\mathcal{S}|}$ the precoding matrix, it has been shown in [248, 249] that the optimal solution of this problem is given by

$$\mathbf{V}^{\text{opt}} = \underbrace{\left(\sum_{\forall i \in \mathcal{S}} \lambda_i^{\text{opt}} \mathbf{h}_i \mathbf{h}_i^H + M\mathbf{I} \right)^{-1} \bar{\mathbf{H}}}_{\triangleq \mathbf{A} = [\mathbf{a}_1, \mathbf{a}_2, \cdots, \mathbf{a}_{|\mathcal{S}|}] \in \mathbb{C}^{M \times |\mathcal{S}|}} \sqrt{\mathbf{P}^{\text{opt}}}, \tag{7.16}$$

where $\bar{\mathbf{H}} = [\mathbf{h}_1, \mathbf{h}_2, \cdots, \mathbf{h}_{|\mathcal{S}|}]$, λ_i^{opt} is the positive unique fixed point of

$$\left(1 + \frac{1}{\gamma_i^\circ}\right) \lambda_i^{\text{opt}} = \frac{1}{\mathbf{h}_i^H \left(\sum_{\forall j \in \mathcal{S}} \lambda_j^{\text{opt}} \mathbf{h}_j \mathbf{h}_j^H + M\mathbf{I} \right)^{-1} \mathbf{h}_i}, \quad \forall i \in \mathcal{S}, \tag{7.17}$$

while $\mathbf{P}^{\text{opt}} = \text{diag}(\mathbf{p})$ with

$$\mathbf{p} = (\sigma_{a,i}^2 + \sigma_{c,i}^2) \mathbf{D}^{-1} \mathbf{1}_{|\mathcal{S}|}, \tag{7.18}$$

and $\mathbf{D}$ is constructed as

$$D_{i,j} = \begin{cases} \frac{1}{\gamma_i^\circ} |\mathbf{h}_i^H \mathbf{a}_j|^2 & \text{if } i = j, \\ -|\mathbf{h}_i^H \mathbf{a}_j|^2 & \text{if } i \neq j. \end{cases} \tag{7.19}$$

7.3.2 Power splitting

Under the PS protocol, the HAP transmits the information signal to the devices while simultaneously energizing them. Specifically, the signal received at each device $i \in \mathcal{S}$ goes through a power splitter, which divides a portion $\rho_i \in [0, 1]$ of the signal power to the EH circuit, and the remaining $1 - \rho_i$ portion of power to the ID circuit. As a result, the EH and ID signal components at the ith device are expressed as

$$y_i^{\text{eh}} = \sqrt{\rho_i} \left(\mathbf{h}_i^H \sum_{\forall j \in \mathcal{S}} \mathbf{q}_j s_j^{\text{swipt}} \right), \tag{7.20}$$

$$y_i^{\text{id}} = \sqrt{1 - \rho_i} \left(\mathbf{h}_i^H \sum_{\forall j \in \mathcal{S}} \mathbf{q}_j s_j^{\text{swipt}} + u_{a,i} \right) + u_{c,i}. \tag{7.21}$$

Again, the noise impact in (7.20) can be ignored. As for the information receive signal, observe that the power of the noise introduced by the RF band to baseband signal conversion $\sigma_{c,i}^2$ remains unchanged; however, differently than in the previous section, herein the antenna noise power $\sigma_{a,i}^2$ is reduced a $\rho_i \times 100\%$ jointly with the information signal power.

Problem Formulation. Based on (7.20) and (7.21), the energy harvested by the ith EH device, and the SINR of the receive information signal, are respectively given by

$$E_i^h = g\left(\rho_i \sum_{j \in \mathcal{S}} |\mathbf{h}_i^H \mathbf{q}_j|^2 \right), \tag{7.22}$$

$$\gamma_i = \frac{(1 - \rho_i)|\mathbf{h}_i^H \mathbf{q}_i|^2}{(1 - \rho_i)\sum_{k \in \mathcal{S} \setminus i} |\mathbf{h}_i^H \mathbf{q}_k|^2 + (1 - \rho_i)\sigma_{a,i}^2 + \sigma_{c,i}^2}. \tag{7.23}$$

Then, the QoS constraints are given by

$$\text{EH:} \qquad E_i^h \geq \xi_i \rightarrow \qquad\qquad \sum_{j \in \mathcal{S}} |\mathbf{h}_i^H \mathbf{q}_j|^2 \geq \frac{g^{-1}(\xi_i)}{\rho_i} \triangleq \xi_i^\circ, \tag{7.24}$$

$$\text{ID:} \qquad \log_2\left(1 + \frac{\gamma_i}{\Upsilon}\right) \geq r_i \rightarrow \qquad \gamma_i \geq \left(2^{r_i} - 1\right)\Upsilon \triangleq \gamma_i^\circ. \tag{7.25}$$

Now, the precoding optimization problem can be written as

$$\textbf{P14:} \quad \min_{\mathbf{q}_i \in \mathbb{C}^{M \times 1}, \forall i} \sum_{\forall i \in \mathcal{S}} ||\mathbf{q}_i||^2 \tag{7.26a}$$

$$\text{s. t.} \quad \text{constraints (7.24), (7.25),} \ \forall i \in \mathcal{S}. \tag{7.26b}$$

Similar to **P10**, the feasibility of **P14** is delimited by (7.6). Moreover, since the charging time is not potentially reduced, the WET-related specific feasibility is superior to that of the TS protocol.

Optimal Solution. As on several previous occasions, herein we apply the SDR technique. To this end, define $\mathbf{Q}_i = \mathbf{q}_i \mathbf{q}_i^H$, thus, $\text{rank}(\mathbf{Q}_i) \leq 1$, $\forall i$. Then, by ignoring such rank-one constraint, the SDR of **P14** is given by

$$\textbf{P15:} \quad \min_{\mathbf{Q}_i \in \mathbb{C}^{M \times M}, \forall i} \sum_{\forall i \in \mathcal{S}} \text{Tr}(\mathbf{Q}_i) \tag{7.27a}$$

$$\text{s. t.} \qquad\qquad \sum_{\forall j \in \mathcal{S}} \text{Tr}(\mathbf{H}_i \mathbf{Q}_j) \geq \xi_i^\circ, \tag{7.27b}$$

$$\frac{1}{\gamma_i^\circ}\text{Tr}(\mathbf{H}_i \mathbf{Q}_i) - \sum_{\forall j \in \mathcal{S} \setminus i} \text{Tr}(\mathbf{H}_i \mathbf{Q}_j) \geq \sigma_{a,i}^2 + \frac{\sigma_{c,i}^2}{1 - \rho_i}, \tag{7.27c}$$

$$\mathbf{Q}_i \succeq 0, \tag{7.27d}$$

where constraints (7.27b)-(7.27d) apply $\forall i \in \mathcal{S}$.

It has been shown in [250] that the optimal solution of **P15**, $\{\mathbf{Q}_i^{\text{opt}}\}$, satisfies the rank-one constraint. Therefore, the optimum precoders can be straightforwardly obtained as the eigenvectors of $\{\mathbf{Q}_i^{\text{opt}}\}$.

7.3.3 TS versus PS

Herein, we illustrate some numerical results on the performance of the TS and PS protocols. Unless stated otherwise, results are obtained after setting $M = |\mathcal{S}| = 8$, $\xi_i = \xi = 70\ \mu$W, $r_i = r = 1$ bpcu, $\sigma_{a,i}^2 = \sigma_{c,i}^2 = \sigma^2 = -90$ dBm, $\Upsilon = 9.8$ dB (assuming that an uncoded quadrature amplitude modulation is employed [226]), and adopting an i.i.d. Rician fading model with LOS factor of 5 dB for the fast fading channels. Moreover, users' channels experience a path-loss of 50 dB, which may correspond to the path-loss experienced at a distance of 18 m when using the log-distance model in (4.21) and (5.2). Finally, we adopt the EH transfer function in (4.25) [100] and set its parameters similar to the setup in Section 6.3.4, producing the result in Figure 6.7.

On the Impact of the TS/PS Coefficient. Figure 7.5 illustrates the average optimum HAP transmit energy for both TS and PS protocols as a function of the WET TS/PS coefficient, τ and ρ, respectively. In the results, the ideal case refers to the scenario where the ID capability is just affected by the noise (there is no external interference) as considered in the analysis and discussions in previous subsections. Observe that TS can provide significant performance gains with respect to PS in such scenarios, i.e., more than 15 dB of power savings. Moreover, since the EH devices are at relatively short distances from the HAP in order to harvest significant amounts of energy, the HAP's transmit energy consumption is dominated by the WET energy expenditure. Thus, the system performance can be seriously affected by adopting a small τ, ρ coefficient. In fact, there is an optimum TS/PS coefficient, which for PS is shown to be close to 1 since the ID process is less sensitive to the choice of ρ as it affects both the numerator and most significant term of the denominator in the SINR expression (7.23). Only $\sigma_{c,i}^2$ is not affected by ρ, but again, the noise impact is very limited in these setups since communication links are short.

Meanwhile, a more practical setup including the influence of a certain external interference is also considered here. Let us denote I_i as the instantaneous interference power experienced by the ith device. Obviously, such values cannot be known in advance at the HAP, thus, the precoder cannot directly use them. It is, however, feasible to acquire the statistics of I_i over time and use these as follows. Let us denote $F_{I_i}(x)$ as the CDF of I_i, then $I_i^\circ = F_{I_i}^{-1}(1 - \varepsilon)$ denotes a power threshold that can be surpassed by the instantaneous interference power with probability ε. Then, by adopting a sufficiently small ε, and plugging the corresponding I_i° into the denominator of the SINR expressions ((7.11) for TS, and (7.23) for PS but first scaled by $1 - \rho$), one can

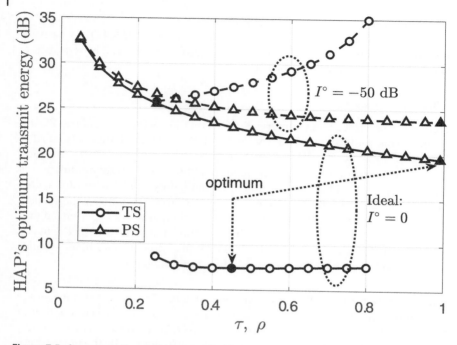

Figure 7.5 Average optimum HAP transmit energy as a function of the WET TS/PS coefficient.

make the precoding resulting from the optimization in Sections 7.3.1 and 7.3.2 more robust since it now includes interference-aware features and corresponding reliability guarantees. For the results in Figure 7.5 considering robustness against external interference, we set $I_i^\circ = I^\circ = -50$ dB. Interestingly, PS now performs better than TS most of the time, and its performance is not seriously degraded with respect to the ideal system without any interference, i.e., just ~ 5 dB of maximum performance gap. Meanwhile, the performance gap between the ideal and non-ideal system is in the order of 20 dB when using TS since the power consumption is now dominated by the WIT transmit power, which is not the case when using PS.

On the Optimal TS/PS Coefficient. The optimal WET coefficient for TS, τ, and PS, ρ, protocols is illustrated in Figure 7.6(a) and Figure 7.7(a) as a function of the path-loss and EH requirement threshold, respectively, while the corresponding average transmit energy of the HAP is depicted in Figure 7.6(b) and Figure 7.7(b).

Interestingly, the optimal TS/PS coefficient is independent of the path-loss, while the corresponding HAP's transmit energy decreases following a power-law (linearly in a log-log scale) as the path-loss increases. Meanwhile, as the EH

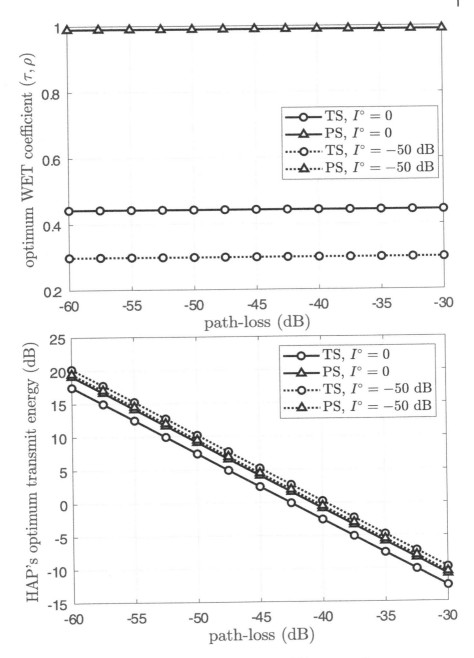

Figure 7.6 (a) Optimum WET TS/PS coefficient (top), and (b) corresponding average optimum HAP transmit energy (bottom), as a function of the path loss.

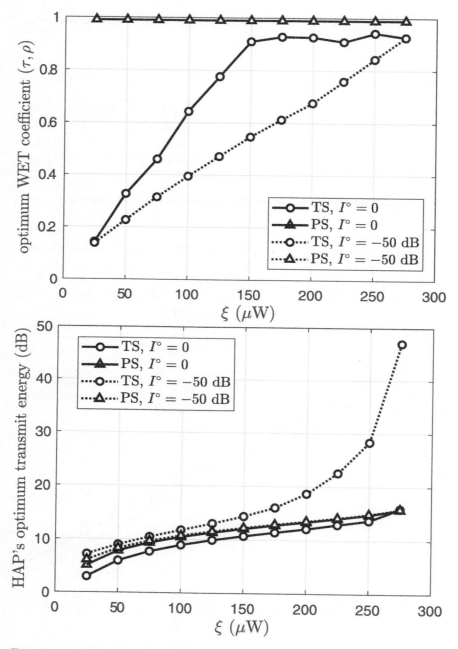

Figure 7.7 (a) Optimum WET TS/PS coefficient (top), and (b) corresponding average optimum HAP transmit energy (bottom), as a function of the EH threshold ξ.

requirements become more stringent, the HAP's transmit energy consumption increases. The consumption can be very abrupt if the TS protocol is adopted and configured for robustness against strong external interference sources, that is $I^\circ = -50$ dB. Finally, while the optimal PS configuration tends to $\rho \to 1$ in most of the cases of practical interest, the optimal TS coefficient does significantly vary, e.g., as a function of the EH threshold and the statistics of the external interference as illustrated in Figure 7.5 and Figure 7.7(a).

7.4 Enablers for Efficient SWIPT

Herein, we discuss some specific promising approaches to boost the performance of SWIPT systems in addition to those discussed in Section 2.4 for general WET-powered IoT networks.

7.4.1 Waveform Optimization

Throughout this book, we have focused on designing efficient strategies to boost the RF energy at the input of the EH circuits with/without additional ID related requirements. However, there is additionally another appealing approach to increase the output DC power level, that is the harvested energy, which lies in designing efficient WET or SWIPT signals (for WPCN and SWIPT systems, respectively). This comes from the fact that the RF-to-DC conversion efficiency is a function of the input waveforms [230, 251–257]. Therefore, traditional Gaussian signaling, which is known to be optimum for pure WIT[1], is not longer appropriate. In the following, we revise the modeling and main approaches for designing efficient waveforms for WET/SWIPT.

Circuit and Signal Modeling. A model that captures input power and shape dependency on output DC power was first derived in [254, 256]. Next, we summarize its main features. Consider the simplified antenna and rectifying circuit shown in Figure 7.8. The lossless antenna is modeled as an equivalent voltage source v_s in series with an impedance R_{ant}, while the rectifier has input impedance R_{in} and is composed of a single series diode followed by a low-pass filter with a load. Note that this is the simplest rectifier configuration, although, the model *per se* is not limited to this, and in fact holds for more general rectifiers with many diodes [255].

Now, assume the RF signal at time instant t impinging on the receive antenna, $y_{rf}(t)$, has an average power P_{rf}. With perfect matching, that is $R_{in} = R_{ant}$, the input voltage of the rectifier can be related to the received signal by

---

1 Gaussian waveform is optimal when the signal is only subject to an average power constraint. Meanwhile, the optimal capacity achieving distribution under both average power and amplitude constraints is discrete with a finite number of mass points for the amplitude, and continuous uniform for the phase [258, 259].

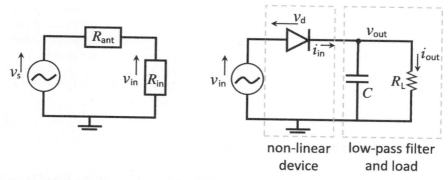

Figure 7.8 (a) Antenna equivalent circuit (left), and (b) a single diode rectifier (right). The rectifier comprises a non-linear device (diode) and a low-pass filter (consisting of a capacitor C and a load R_L). *Source:* [254, Fig. 1].

$$v_{in}(t) = y_{rf}(t)\sqrt{R_{ant}}. \tag{7.28}$$

Then, by neglecting the diode's resistance (ideal diode), the current $i_d(t)$ relates to the voltage drop across the diode $v_d(t) = v_{in}(t) - v_{out}(t)$ as

$$i_d(t) = i_s\left(e^{\frac{v_d(t)}{\eta v_0}} - 1\right), \tag{7.29}$$

where i_s is the reverse bias saturation current, v_0 is the thermal voltage, and η is the ideality factor. Now, the DC current delivered to the load and the harvested DC power are given by

$$i_{out} = \mathbb{E}[i_d(t)], \qquad P_{dc} = i_{out}^2 R_L, \tag{7.30}$$

respectively. Note that the expectation operator not only takes the DC component of the diode current $i_d(t)$ but also averages over the potential randomness carried by the input signal $y_{rf}(t)$.[2] This randomness impacts the diode current $i_d(t)$ and the amount of harvested power, which is captured in the model by taking an expectation over the distribution of the input symbols.

Because of the intricate dependence of $i_d(t)$ and $y_{rf}(t)$ through (7.29) and (7.28), it is not often feasible to work directly with the above model without some further assumptions and simplifications. For instance, using Taylor expansion around a quiescent operating point $v_d(t) = \bar{v}$, (7.29) can be written as [142]

---

2 In SWIPT systems, $y_{rf}(t)$ may potentially carry also information, and may be therefore changing at every symbol period due to the randomness of the input symbols.

$$i_d(t) = i_s \sum_{k=0}^{\infty} c_k (v_d(t) - \bar{v})^k,$$ (7.31)

where

$$c_k = \begin{cases} i_s \left(e^{\frac{\bar{v}}{\eta v_0}} - 1 \right), & \text{for } k = 0, \\ \frac{i_s}{k!(\eta v_0)^k} e^{\frac{\bar{v}}{\eta v_0}}, & \text{for } k = 1, 2, \cdots. \end{cases}$$ (7.32)

Now, assume a steady-state response and an ideal rectification such that $v_{\text{out}}(t)$ is at constant DC level v_{out}. Then, by choosing $\bar{v} = \mathbb{E}[v_d(t)] = -v_{\text{out}}$, using (7.28), and truncating the Taylor expansion at the k'th order, (7.31) transforms to

$$i_d(t) = \sum_{k=0}^{\infty} c_k v_{\text{in}}^k = \sum_{k=0}^{\infty} c_k R_{\text{ant}}^{k/2} y_{\text{rf}}(t)^k \approx \sum_{k \text{ even}}^{k'} c_k R_{\text{ant}}^{k/2} y_{\text{rf}}(t)^k,$$ (7.33)

where $k' \geq 2$ is an even integer, thus i_{out} approximates to

$$i_{\text{out}} \approx \sum_{k \text{ even}}^{k'} c_k R_{\text{ant}}^{k/2} \mathbb{E}[y_{\text{rf}}(t)^k].$$ (7.34)

However, since each c_k is a function of $\bar{v} = -v_{\text{out}} = -R_L i_{\text{out}}$, it is still difficult to express i_{out} explicitly as a function of $y_{\text{rf}}(t)$. Fortunately, maximizing i_{out} in (7.34) subject to an RF transmit power constraint, and consequently P_{dc}, is known to be equivalent, from a signal waveform optimization perspective, to maximizing [254–256]

$$i'_{\text{dc}} = \sum_{k \text{ even}, k \geq 2}^{k'} c'_k \mathbb{E}[y_{\text{rf}}(t)^k],$$ (7.35)

where $c'_k = i_s R_{\text{ant}}^{k/2} (\eta v_0)^{-k} / k!$. Therefore, a WET/SWIPT waveform optimization framework can directly benefit from (7.35) as herein i'_{dc} is not only influenced by the RF receive power but also by the signal shape. Observe that the waveform-independent model used repeatedly throughout this book's chapters is obtained by truncating (7.35) at order 2. However, whenever the higher order terms of (7.35) are found to be non-negligible, they should be taken into account for better accuracy, and in fact, they critically influence the waveform optimization as illustrated in [230, 251–257].

Optimization Goals. In a WPCN setup, or SWIPT system with TS receivers, WET and WIT waveforms can be independently designed. Therefore, the optimum WIT waveform is not affected by the EH circuit particularities discussed above, and a capacity-achieving/approaching WIT waveform can be adopted. Meanwhile, the optimum WET waveform is the one that maximizes (7.35). In [256], the authors show that modulated waveforms are beneficial for single-carrier energy transmissions, but detrimental in multi-carrier transmissions.

The story is different in a SWIPT setup with PS receivers, where the same waveform carries both information and energy. In the simplest single-input single-output (SISO) single frequency sub-band scenario, the receive RF signal can be written as a function of the transmit signal x as

$$y = hx + n, \tag{7.36}$$

where h is the channel coefficient, and n denotes the AWGN. Observe that y in (7.36) relates to $y_{\rm rf}$ in (7.28) by $y_{\rm rf} = \sqrt{1 - \rho} y$. Meanwhile, the receive signal at the input of the ID circuit is $y_{\rm id} = \sqrt{\rho} y$.

By dropping the Gaussian signaling assumption, which is only optimum in pure WIT setups with transmit average power constraints, one can no longer adopt Shannon capacity formula to characterize the achievable rate. Instead, the mutual information between the channel input x and the channel output $y_{\rm id}$,

$$I(x, y_{\rm id}) = \int \int f_{X,Y_{\rm id}}(x, y_{\rm id}) \ln \left(\frac{f_{X,Y_{\rm id}}(x, y_{\rm id})}{f_X(x) f_{Y_{\rm id}}(y_{\rm id})} \right) \mathrm{d}x \mathrm{d}y_{\rm id}, \tag{7.37}$$

needs to be considered.

Using this framework the authors of [256] provide evidence that a multi-carrier unmodulated waveform is useful to enlarge the rate-energy region of SWIPT, while a combination of PS and TS is in general the most suitable strategy. Moreover, a non-zero mean Gaussian distributed channel input is shown to outperform the conventional capacity-achieving zero-mean Gaussian distributed channel input in multi-carrier transmissions.

7.4.2 Multicarrier SWIPT

Orthogonal frequency division multiplexing (OFDM) is a widely used multicarrier technique in various wireless standards, including 5G. Thus, several existing papers had already studied the performance of OFDM-based SWIPT systems, for example [253, 260–264].

Zhou *et al.* [260] studied the weighted sum-rate performance of multi-user SWIPT systems subject to per-user EH constraints under TDMA and orthogonal frequency division multiple access (OFDMA). The TS protocol is used in the TDMA setup in such a way that the ID receiver of a particular user operates in a scheduled time slot, whereas the energy receiver operates in all other time periods. Meanwhile, the PS protocol is applied at each receiver in the OFDMA setup considering all subcarriers share the same PS coefficient at each receiver. The authors showed that the PS scheme may outperform TS when the EH requirements are sufficiently stringent. Meanwhile, a joint design of resource allocation and PS ratios was proposed in [261] for multi-group multi-cast OFDM systems. The proposal consists of a subcarrier allocation method followed by a power allocation and PS ratios optimization procedure. On the other hand, a MIMO OFDM network in the presence of a passive eavesdropper is considered in [262]. Therein, both WET at the receiver and the use of hybrid artificial noise at the transmitter are taken into account. Moreover, the PS SWIPT technique is utilized, while the cyclic prefix of each OFDM block is used for the purpose of EH at the legal receiver. The authors proposed a trade-off problem to maximize the secrecy rate of the network while satisfying EH requirements. The OFDM signal structure is also leveraged in [264], where the authors proposed and analyzed the functionality of a novel selective OFDM receiver architecture. Therein, the cyclic prefix of the OFDM signal, and a selectively chosen fraction of the received signal, are used for EH. Selective OFDM transmission is found to provide improved EH and sustainability.

In terms of waveform optimization, in [253] the authors investigated a SWIPT scenario with a superposition of multi-sine and OFDM transmit waveforms. The receive devices are equipped with PS circuits where the ID blocks are capable of canceling the multi-sine waveforms. The waveform design adapts to the CSI to optimize the SWIPT rate-energy region. Moreover, a scheme where the SWIPT transmitter superimposes a DC level on the OFDM signal to generate an OFDM-DC waveform was proposed in [263]. Therein, the OFDM signal component was used for information transmission, and the DC signal is introduced for improved WET capabilities. The authors discussed the benefits of such an approach and presented some initial evaluations which promote its potentials.

Finally, it is worth highlighting that most of the state-of-the-art research on SWIPT has focused on narrowband receivers for EH and ID purposes, while it would be interesting to investigate wideband SWIPT receiver architectures. One of the promising research directions in this area is to investigate suitable techniques to enable the utilization of all the unused carriers for EH and the information bearing carrier for ID [265]. Therefore, the SWIPT receiver should be capable of working in multiple frequency bands and probably incorporate some low-complex intelligence that allows, for instance, tuning appropriate frequency bands and autonomously configure receive parameters for optimum performance.

7.4.3 Cooperative Relaying

In wireless networks, the transmit and receive nodes may not be sufficiently close to each other or their channel may be obstructed, hence, LOS communication, WET, or SWIPT, may be infeasible. To overcome this, idle nodes can be used to create intermediate feasible hops and assist them, in what is called as cooperative relaying.

Recently, SWIPT cooperative networks have been extensively studied, for instance in [266–276]. In [267], the authors maximized the energy efficiency of SWIPT amplify and forward (AF) MIMO relay networks by jointly designing the source and relay precoding matrices. Meanwhile, a robust energy efficiency optimization problem for a SWIPT MIMO two-way relay system under imperfect CSI was formulated and solved in [275]. With a focus on TS based SWIPT relay systems, the power allocation and flexible TS ratios to maximize the average system throughput were proposed in [266]. Moreover, the throughput and delay metrics were investigated in [165] for a decode and forward (DF) relaying mechanism in a TS SWIPT system operating with short block lengths. Therein, the authors evidenced that the possibility of meeting stringent reliability and latency constraints increases by decreasing the message length and/or for relay positions closer to the source node. Power allocation schemes have also been studied in different SWIPT relay networks to improve the throughput, for example in [268, 269]. Furthermore, relay selection has been explored in multiple-relay scenarios [270–272, 276]. In consideration of the available CSI, different relay selection schemes, that is, first-link relay selection, random relay selection, and distance-based relay selection, lead to various performance. In [273], NOMA techniques were adopted in SWIPT cooperative networks to improve the system energy efficiency Moreover, an optimal resource allocation scheme has been developed in [274] to maximize the energy efficiency of SWIPT two-way DF relays, where the transmit power and PS ratios are jointly optimized.

However, further research is still needed, especially to determine efficient channel codes that improve the *net* coding gain, which takes into account not only the throughput coding gain but also the power consumption cost. As a first step, the power consumption of available channel codes such as polar codes and low-density parity-check codes need to be investigated in order to determine their practical suitability for SWIPT systems. Finally, the mobility of relay nodes and its effect on EH also need to be addressed in future research.

7.4.4 Interference Exploitation

The interference phenomenon has been traditionally considered a foe, a problem to be avoided or at least partially mitigated, in wireless communication systems. However, more recently, some innovative ways of using interference are emerging to either improve the reliability, security, and the achievable rate of wireless networks [277]. In the specific case of SWIPT systems, it must be noted that although interference alone (without any potential exploitation technique) may be detrimental for the ID component, it is on the other hand beneficial for the EH process. Therefore, a proper balance between ID and EH functionality/performance is required to facilitate efficient SWIPT.

SWIPT under interference has been broadly studied in multi-user systems, for example in [277–284]. Interestingly, authors in [279, 282] proposed a multi-user symbol-level precoding approach to address the multi-user interference through jointly utilizing the CSI and data information. Therein, the interference among the data streams is converted under certain conditions to a useful signal that potentially improves the SINR of the downlink channel. Some efficient beamforming strategies have also been proposed in [282] to exploit the concept of mutual coupling between the transmit antennas by using tunable antenna loads. Additionally, in [280], the authors proposed an opportunistic communication scheme for interference alignment based on antenna and user selection to achieve efficient SWIPT. Meanwhile, constructive interference mechanisms were investigated in [283] to improve the performance of both the EH and ID in a MISO downlink system.

Finally, observe that FD SWIPT schemes allow self-energy recycling in multi-antenna nodes, which benefit them with a secondary energy source when the receive antennas are used for EH [157].

7.4.5 Artificial Intelligence

WET and SWIPT systems are subject to non-linear responses that are often model-intractable unless too strong assumptions are adopted. The lack of tractable mathematical models of the rectenna, and more generally of the entire WET chain (transmitter-receiver, including high power amplifier), as well as algorithms to solve WET/SWIPT signal/system optimizations in general settings, is a bottleneck towards efficient designs [257].

AI techniques can be used to circumvent these modeling and algorithmic challenges. They can even take advantage of large training datasets that can be created from circuit simulations and prototyping. The current drawbacks lie in the lack of performance guarantees on the accuracy of the solutions (which may, however, be somewhat overcome by using risk-sensitive approaches), and the lack of human-interpretability. This is why a hybrid or integrated approach combining the best of the model-based [76, 77, 85, 142, 231, 256] and AI-based [285–287] WET/SWIPT optimization may be more attractive, and example of it is the so-called *physics learning* [288].

7.5 Final Considerations

In this chapter, we studied SWIPT scenarios. Specifically, we discussed the main SWIPT receivers, namely, separate and co-located receivers. In the former, EH and ID circuit blocks are fed by independent antennas, while the same set of antennas is used for the purposes of both EH and ID in the latter. We illustrated the performance of a representative scenario for each of these. In all cases, the receivers are subject to EH and ID constraints, and the system performance is measured in terms of HAP transmit energy consumption.

For the scenario with separate SWIPT receivers, where a HAP simultaneously powers a set of EH devices and transmits information to a set of ID devices, we found the optimum precoding vectors. Interestingly, results evidenced that no more than one WET signal is needed for optimum performance when the ID receivers are able to perfectly cancel its effect using SIC. Meanwhile, under imperfect WET SIC at the ID receivers, more than one WET signal may be needed. Fortunately, the system performance does not degrade significantly as the corresponding SIC imperfection parameter increases. However, our analyses are based on the assumption that such a parameter is known in advance, which may not hold in practice, thus demanding further studies. Other important observations are that the HAP's energy requirements increase 1. exponentially with the number of served devices, 2. proportional to the EH requirements of the device, and 3. slowly with the SINR requirements of the device.

For the scenario with co-located SWIPT receivers, where a HAP simultaneously powers and transmits information to a set of devices with both EH and ID capabilities, we evaluated the performance of TS and PS protocols with optimum transmit precoding. Results showed that TS can easily outperform PS in interference-free scenarios, while PS can make the system robust against external interference impairments without significantly increasing the HAP's transmit energy requirements. In addition, we investigated the optimal TS/PS coefficient to be configured in the receive network, and showed that it is independent of the channel's path-loss.

Finally, we concluded the chapter with an overview of some potential techniques to enable efficient SWIPT in future networks. Our discussions focused on WET/SWIPT waveform optimization, multi-carrier transmissions, cooperative relaying, interference exploitation, and AI mechanisms. We provided a state of the art revision on these, discussed their main innovative aspects, and highlighted some key research directions.

8

Final Notes

ICT industry and academy are relentlessly stepping up efforts to realize a data-driven sustainable society, which demands challenging near-instant, zero-energy, unlimited wireless connectivity. Massive IoT network deployments are on the way to being the main connectivity backbone of such a future society. However, there is a major concern related to the lack of energy efficient solutions for powering and keeping uninterrupted operation of the massive number of devices already emerging and that are subject to increasingly stringent QoS requirements. In this regard, EH techniques constitute an attractive solution as they allow batteries to be recharged externally and/or to not require replacement, which would be otherwise not only environmentally-unfriendly and costly but also impossible in hazardous environments, building structures or the human body. We foresee EH as a key component of any future massive low-power IoT network.

This book has focused specifically on the RF–EH technology, which is shown to be a stronger candidate for powering many low-power massive IoT use cases than EH technologies based on other energy sources.

Next, we summarize the most relevant ideas and findings discussed in Chapters 1–7, and highlight relevant future research directions, some of which have already been suggested in previous chapters.

8.1 Summary

Chapter 1. A massive number of IoT devices will be continuously deployed in the coming years at an exponential pace. Some selected representative use-cases and application scenarios are: smart-metering, E-health life sign monitoring and wearables, smart-agriculture, transport and logistics, and industry, with diverse requirements in terms of energy efficiency, scalability, coverage, device cost, data rates, and security. To face this, massive IoT technological

Wireless RF Energy Transfer in the Massive IoT Era: Towards Sustainable Zero-energy Networks.
First Edition. Onel Alcaraz López and Hirley Alves.

solutions, nowadays mostly in the hands of LoRa, SigFox, LTE-M, NB-IoT and 5G mMTC, will experience a multi-fold increase in the next years while taking market share from the legacy network. This will be supported by the adoption of promising sustainability-promoting technologies such as IRS, TinyML, intelligent access and resource management, DAS, efficient short-length coding, CSI-free/limited solutions, zero-energy devices, EH and backscatter communications.

Chapter 2. EH technology based on RF energy is a strong candidate for powering future low-power massive IoT deployments. The RF transmitters, medium, and EH nodes constitute the three key components of a RF-EH system. According to the nature of the RF transmitters, RF-EH technologies are classified as ambient RF EH, or dedicated RF EH (mainly referred to as WET in this book). Although there exist several analytical models to mimic the input–output relation of RF–EH circuits, there is no one-size-fits-all model because of the trade-off between accuracy and tractability. Meanwhile, the EH circuits are still designed with a specific application/scenario in mind, thus, their circuit features vary accordingly but are mostly inflexible. Since the IoT paradigm intrinsically includes WIT, RF-EH appears naturally combined with the former in an infrastructure-based/-less WPCN or SWIPT architecture. The use of green energy sources at the RF transmitters, CSI-limited EB and scheduling schemes, DAS, and lightweight DLT mechanisms constitute promising techniques for enabling sustainable RF-WET in the IoT era.

Chapter 3. Ambient RF–EH techniques exploit readily available environmental RF energy. There exist three key energy usage protocols at the receiver side, namely, HU, HSU and HUS. The most widely adopted protocol, HU, performs the worst since no energy saving is allowed, while the relative performance of HUS and HSU protocols depend on the battery impairments. The ambient RF power density in a certain area is significantly small. Thus, high-gain multi-/tunable-band antennas, high-gain multi-directional rectenna circuits, and/or large harvesting surfaces are desirable. In some scenarios, relying completely on RF–EH sources is not possible because of the low harvested energy and energy arrival rate uncertainty, thus, hybrid solutions are needed. Ambient RF-EH technology is mainly appealing for supporting ultra-low power IoT applications, and its feasibility is commonly assessed via on-site RF measurements, which is costly. The latter can be alleviated by partially using mathematical tools that provide an intuitive and tractable framework for statistical feasibility analysis. In fact, the considerable performance improvements brought by multi-band and tunable-band harvesters can be easily validated via a stochastic-geometry based study.

Chapter 4. WET refers to on-purpose energy transmissions from dedicated energy sources to EH devices, which no longer strictly demand multi-/tunable-band receivers. Massive WET refers to WET to a massive number of low-power MTDs. Novel CSI-limited/-free EB and scheduling mechanisms are key to supporting massive WET since the gains from traditional full and perfect CSI-based schemes may not compensate the energy consumed during the CSI acquisition procedures. Specifically, an average-CSI EB often leads to near optimum performance results since it is mainly influenced by the LOS channel component, which is usually strong in WET setups. However, as the number of EH devices increases, CSI-free techniques are more desirable. The performance of state-of-the-art CSI-free solutions can be improved by exploiting devices' positioning information, which may be available in the case of static or quasi-static IoT deployments, devices' battery state information and/or associated QoS requirements, and by incorporating rotary capabilities into the PBs.

Chapter 5. Multi-PB systems are required for banning RF energy delivery blind spots in relatively-wide EH areas. An efficient deployment of PBs must exploit devices' positioning, clustering information, and/or deployment area features. Proper power control and multi-antenna EB mechanisms can also significantly improve WET efficiency. The deployment optimization of PBs and EB design are in general two problems that cannot be coupled since the latter commonly requires some kind of time-varying CSI. Centralized EB can significantly outperform distributed EB under ideal conditions (that is, perfect, instantaneous, and cost-free CSI-acquisition, and high-capacity control channels between the PBs and a central unit that runs the optimization); however, their relative performance is still unclear under non-ideal conditions. Meanwhile, whenever multi-antenna PBs are deployed to operate with a CSI-free mechanism, the deployment optimization can be directly influenced by its choice. Single-/multi-antenna PBs deployments under QoS requirements are the most cost-friendly when PBs acquisition costs increase linearly/logarithmically with the number of antennas.

Chapter 6. In a WPCN, PBs or HAPs perform in a first phase WET to EH devices, which then use the harvested energy to transmit their own data in a second phase. At the EH device side, adopting an HTT protocol is sub-optimal. Instead, a power control mechanism, that allows saving part of the harvested energy for future use, leads to better performance results. Based on CSI information, a device can set a transmit power threshold that allows communication with a target decoding error probability ε_0. In the high data rate regime, i.e., $r \to \infty$, the transmit power threshold is proportional to $e^{\frac{1}{\sqrt{n}}Q^{-1}(\varepsilon_0)}(2^r - 1)$, where n is the information block length. There exists a non-trivial optimum transmit threshold configuration since the smaller ε_0, the greater the chances of energy outage; while the larger ε_0, the greater the chances of decoding error. In a multi-user multi-antenna HTT WPCN with per-user rate constraints, TDMA outperforms SDMA only when implemented at a HAP with a small number of antennas. In general, system performance worsens as the devices' activation probability and/or devices' circuit power consumption increase, which may be counteracted by equipping the HAP with more antennas.

Chapter 7. SWIPT refers to scenarios where WET and WIT occur simultaneously in the same link direction. When a HAP simultaneously powers a set of EH devices and transmit information to another set of ID devices (separate SWIPT receivers), no more than one WET signal is needed for optimum performance as long as such a signal can be perfectly canceled at the ID receivers. In such setups, the HAP's energy requirements increase exponentially with the number of served devices, proportional to the devices' EH requirements, and slowly with the devices' SINR requirements. Meanwhile, when the same antenna is used for both EH and ID purposes at each device (co-located SWIPT receiver), WET and WIT can be scheduled via TS or PS protocol. TS can easily outperform PS in interference-free scenarios, while PS can make the system robust against external interference impairments without significantly increasing the HAP's transmit energy consumption. SWIPT performance can be significantly improved by exploiting efficient waveforms, multicarrier transmissions, cooperative relaying, interference exploitation, and/or AI mechanisms.

8.2 Future Research Directions

Efficient Training via Evolved MAC. Traditional training for CSI acquisition may be prohibitive in massive low-power IoT networks, which is why several CSI-limited/-free solutions have been recently proposed. Still, channel training may be feasible if more evolved scheduling and MAC mechanisms are realized. In general, a basic TDMA, FDMA, or even SDMA scheme, is not appropriate, and much more evolved MAC schemes are required to account for the different (and variable) energy requirements of the EH devices, and the QoS demands of both data transmission and energy transfer in WPCNs. Transmit power, duration, and frequency bands for training need to be carefully designed to minimize the energy spent. A proper MAC orchestration is necessary to avoid pilot collisions and consequent extra energy expenditure in the collision resolution phase.

For instance, consider the following simple scheduling mechanism. Allow only the set of devices with low-energy levels, say a battery level below a threshold ξ, to execute channel training procedures. The greater ξ is, the number of devices competing for the spectrum resources becomes greater, and more energy demands may be simultaneously satisfied; while the smaller ξ, the greater the chances of under-exploiting the spectrum resources and the chances of energy outage, but training becomes less energy-consuming. Therefore, these trade-offs lead to the existence of an optimum ξ, which raises several research questions such as: *How great are the gains that can be obtained with this approach? Under which conditions are CSI-limited/free solutions preferable? What about hybrid mechanisms?* AI mechanisms can also be leveraged to track/predict the energy consumption profile of the devices and act accordingly.

Radio Stripes-enabled High-power WET. Future WET systems will undoubtedly benefit from novel kinds of DAS deployments such as cost-efficient radio stripes systems. In indoor environments, high energy delivery can be realized, for example to energize more power-hungry device such as smart watches and video-game console controllers, due to the potentially massive number of transmit antennas that may be available and the short-range links. Future research here can focus on optimized resource allocation schemes, circuit implementations, prototypes, and efficient distributed processing architectures to avoid costly signaling between the antenna elements.

Reprogrammable Medium. WET propagation medium can be conveniently "influenced" via the strategic deployment of smart reflective-arrays and reconfigurable meta-surfaces. IRS are beneficial to either compensate the significant path losses thanks to their large aperture arrays, or to provide alternative WET links to obstructed direct paths. Towards the massive adoption of IRS technology, some potential research directions lie on realizing novel jointly-designed passive reflect EB of IRS and active EB of PBs/HAPs, and efficient CSI-limited/free schemes. Novel mechanisms for partitioning IRS elements set such that each partition is responsible for reflecting the arriving signal to a certain device or group of devices may be also appealing.

mmWave WET. Exploiting the mmWave spectrum for WET may be advantageous because of 1. the large bandwidths that remain unexploited in such spectrum regions; 2. mmWave signals propagation is more directive and of shorter-range which is favorable for spectrum re-use in small cells; 3. the shorter wavelengths allow antenna sizes to be reduced, which translates to either smaller form-factors and/or to being able to pack more transmit/receive antennas; and 4. integration/coexistence of WET and WIT may be more easily promoted since interference issues are considerably relaxed. Moreover, since LOS channels are characteristic of many WET applications and are mostly influenced by the antenna array topology and network geometry, CSI-based EB may be avoided (at least partially) by exploiting accurate devices' positioning information. However, future research should address the problem of imperfect beam alignment, and how to overcome NLOS, or even severe signal attenuation when operating with limited CSI, which may require exploiting multi-/hybrid-RATs.

DLT-enabled Energy Trading. DLT-based technologies, such as Blockchain and Holochain DLT, are appealing to enable secure energy trading between nodes. Future research efforts must be around mitigating the large communication overhead of nowadays protocols and efficiently handling massive two-way connections, to adapt to massive low-power IoT deployments. In addition, powering the devices that actually paid for the service must be extremely precise in space, time (and other domains), for which novel and efficient EB designs are necessary, for instance, to benefit the legitimate EH devices while providing negligible energy to the non-authorized ones.

AI at PBs and Devices. Collaborative and distributed AI at the devices and the network is foreseen as a key enabler for critical low-power IoT networks. Since WET-powered systems are subject to non-linear responses that are often model-intractable, it is worth investigating novel AI techniques at the network side to circumvent the lack of tractable mathematical models of the rectenna, and more generally of the entire WET chain (transmitter-receiver, including high power amplifier), and to address WET/SWIPT signal/system optimizations. These AI-methods can be strengthened by incorporating physics models, thus resulting in hybrid model/AI-based mechanisms more prone to provide QoS guarantees and better interpretability. Moreover, future research must explicitly consider the energy consumption of AI algorithms, and their sustainability-supporting capabilities. How to minimize the required training data, and how to decrease the complexity of algorithms is worth investigating. Observe that AI mechanisms demanding minimum communication overhead are desirable at the network side, while developing TinyAI blocks is key at the device's side. Novel TinyAI mechanisms may provide a solution to the problem of how to leverage multiple receive antennas (for receive beamforming) in ambient RF–EH. Note that since incoming RF transmissions are not dedicated in these scenarios, CSI cannot be acquired, at least in the standard way using pilot sequences.

Ambient RF–EH Circuits and RF Measurement Campaigns. On the hardware and circuit design front, research efforts can be focused on the prototyping of efficient tunable-band harvesters, high-gain multi-directional rectennas, and large and thin EH surfaces that can be seamlessly integrated into everyday objects, for ambient RF–EH applications. On the other hand, the number of publicly available results from RF measurement campaigns is rather limited nowadays, especially in recent years. This, and the fact that results from RF measuring campaigns may quickly become obsolete as communication networks tend to densify while new technologies are periodically introduced, suggest the need for many more, and continuous, RF measurement campaign efforts and related analysis in the coming years.

Optimum PB/HAP Rotary Mechanisms. Incorporating rotary capabilities into PBs promises significant performance gains. Exploring analytically and algorithmically the PB rotation optimization under full or limited CSI, or even based on device feedback, constitutes an interesting research direction. Moreover, the performance under different antenna array architectures and non-linear EH models, together with novel rotary mechanisms for centrally/distributively-coordinated multi-PB setups need to be investigated. Finally, how WIT should be scheduled and optimized in WPCN or SWIPT setups with rotary PBs/HAPs is a completely open question.

Positioning-aware CSI-free WET Schemes. State-of-the-art CSI-free WET schemes are mostly blind in the sense that they exploit little or no information for performance improvements. However, positioning information, which may be available in the case of static or quasi-static IoT deployments, could be used to improve WET efficiency. For instance, CSI-free precoders can be designed to point relatively-wide energy beams in the direction of clustered IoT devices, which opens research directions in: 1. clustering algorithms favoring the angular domain; 2. power allocation mechanisms that give preference to clusters with the farthest devices; and 3. per-cluster antenna selection mechanisms (since the greater the number of antennas used to power certain cluster, the narrower the associated energy beam). Moreover, positioning information is inexact in practice, and the position accuracy must be taken into account. Finally, efficient designs may also take advantage (when available) of the devices' battery state information and QoS requirements to dynamically reduce the set of devices to be served, and optimize the network performance.

EB Exploiting Heterogeneous CSI Availability. In heterogeneous deployments, a PB may need to simultaneously power devices with different energy demands and for which CSI acquisition procedures are, or are not, affordable. In the latter case, some statistical CSI may be at hand. Efficient EB designs need to consider the resulting network heterogeneity.

Multi-PB/HAP System Optimization. In a multi-PB setup, centralized EB outperforms distributed EB under ideal conditions, say, perfect, instantaneous, and cost-free CSI acquisition, and high-capacity control channels between PBs and the central unit. However, their relative performance is still unclear in practical scenarios. The performance degradation of centralized EB in non-ideal system conditions needs to be further investigated. Moreover, on-line AI-enabled trajectory optimization for moving/flying PBs potentially exploiting full/partial/limited feedback from the IoT devices, for example, battery state information, is worth investigating. Finally, a joint waveform and precoding design for multi-PB/HAP systems constitutes another interesting research direction.

Green WET. Energy resources at PBs/HAPs powered by *in-situ* green energy, from either ambient or dedicated EH, are intrinsically limited or volatile. Thus, efficient distributed energy scheduling, cooperation, and usage protocols are needed to support network-wide WET from PBs/HAPs to IoT devices.

Multi-carrier SWIPT. Most of the state-of-the-art research on SWIPT has focused on narrowband receivers for EH and ID purposes while it would be interesting to investigate wideband SWIPT receiver architectures. One of the promising research directions in this area is to investigate suitable techniques to enable the utilization of all the unused carriers for EH and the information bearing carrier for ID. Therefore, the SWIPT receiver should be capable of working in multiple frequency bands and probably incorporate some TinyAI blocks that allow, for example, tuning appropriate frequency bands and autonomously configuring receive parameters for optimum performance.

Low-power Channel Codes for WPCN/SWIPT Systems. In EH IoT networks, the power consumption is a critical issue. Thus, channel codes that improve the *net* coding gain, which takes into account not only the throughput coding gain but also the power consumption cost, must be adopted in cooperative WPCN/SWIPT networks. As a first step, the power consumption of available channel codes such as polar codes and low-density parity-check codes need to be investigated in order to determine their practical suitability for WPCN/SWIPT systems.

Complying with EMF Regulations. Wireless transmitters are subject to strict regulations on the level of RF radiation to prevent potential harm to humans and the environment. These regulations minimize the potential biological effects (for example, tissue heating, metabolic changes in the brain, carcinogenic effects) linked to RF radiation. Two widely adopted measures/regulations on RF exposure are the specific absorption rate (SAR), and the maximum permissible exposure (MPE). The former measures the absorbed power from fields between 100 kHz and 10 GHz in a unit mass of human tissue by using units of W/kg, and dominates the RF exposure for very short distances (less than few tens of centimeters). Meanwhile, the latter measures the RF EMF radiation in units of W/m^2, and becomes relevant at larger distances and as the frequency grows above 4 GHz. Future innovative proposals addressing the research directions identified in previous items need to comply with these RF radiation regulations, which impose strict optimization constraints.

Appendix A: A Brief Overview on Finite Block Length Coding

Traditional communication systems are often designed and optimized using the notion of channel capacity, which is a reasonable benchmark for systems using long block transmissions. However, many IoT applications and mMTC are often characterized by short packets, periodic data traffic, and above all, by a massive number of deployed devices. As a result, the same assumptions in terms of channel capacity cannot be directly applied to short block length messages [289]. In this context, new theoreticcal results related to the performance of short block length communication systems have shown that the achievable rate depends not only on the channel quality of the communication link, but that it is also a function of the actual block length and error probability tolerable at the receiver [75, 289, 290].

In light of these new results, the underlying channel models need to be revisited for IoT specific environments. Therefore, it is possible to design new protocols that account for such short messages and then optimize coding strategies, as well as rate adaptation and power allocation policies, e.g., as in [76, 77, 164, 165, 237, 291–293]. The literature is growing since Poliyansky's seminal paper [75]. The surveys in [289, 290] provide an account of the key findings in the recent years. All in all, these results are fundamental to understanding the trade-offs between message payload and size (block length), and, therefore, reliability, latency, and data rate.

A.1 Finite Block Length Model

Assume a source wants to communicate with a destination. The source, an MTD or IoT device, conveys a message composed of a payload of b bits. This message is overall short, i.e., spans over a limited number n of channel uses, and, therefore, requires analysis with a new set of information-theoretical tools [75, 289, 290].

The encoder maps the sequence of b information bits to a sequence of information symbols of length n, which are then transmitted over the wireless channel. Then, the decoder's job is to guess the information bits that were transmitted. When the decoder guesses incorrectly, an error is declared. Mathematically, the encoding and decoding translates as follows.

Wireless RF Energy Transfer in the Massive IoT Era: Towards Sustainable Zero-energy Networks.
First Edition. Onel Alcaraz López and Hirley Alves.
© 2022 by The Institute of Electrical and Electronics Engineers, Inc. Published 2022 by John Wiley & Sons, Inc. Companion website: www.wiley.com/go/Alves/WirelessEnergyTransfer

The source transmits its message $\iota \in \{1, \cdots, L\}$ using a (n, L, P, ε) code, where n is the block length, L is the codebook size, P is the power constraint, and ε is the maximum error probability constraint. Thus, the (n, L, P, ε) code comprises:

- An encoder $\Upsilon : \{1, \cdots, L\} \mapsto \mathbb{C}^n$, which maps the message ι into a length-n codeword $x_\iota \in \{x_1, \cdots, x_L\}$ satisfying the power constraint

$$\frac{1}{n} \|x_j\|^2 \leq P, \quad \forall j. \tag{A.1}$$

Observe that in the context of Chapter 6, the available power P is a function of the harvested energy.

- A decoder $\Lambda : \mathbb{C}^n \mapsto \{1, \cdots, L\}$ satisfying the maximum error probability constraint

$$\mathbb{E}\left[\max_{\forall j} \mathbb{P}\left[\Lambda(y) \neq J | J = j \right] \right] \leq \varepsilon, \tag{A.2}$$

where y denotes the channel output induced by the transmitted codeword at the end of each transmission.

Notice that the rate r is defined as the fraction $\frac{b}{n}$ of information bits to the number of transmitted symbols. Ideally, we attempt to design the code such that r is as large as possible, while the error probability is as small as possible. Bearing this in mind, we can define the maximum channel coding rate $r^*(n, L, P, \varepsilon)$ as the largest rate $\frac{\log_2 L}{n}$ (measured in bits per channel use – bpcu) for which there exists a code (n, L, P, ε), which mathematically translates to

$$r^*(n, L, P, \varepsilon) = \sup \left\{ \frac{\log_2 L}{n} : \exists (n, L, P, \varepsilon) \text{ code} \right\} \text{ (bpcu)}. \tag{A.3}$$

Then, P, b, n and ε are related according to (A.3). Polyanskiy *et al.* [75] provide an accurate characterization of the trade-off between these parameters in AWGN channels as follows

$$nr = b \overset{(a)}{\approx} nC(\gamma) - \sqrt{nV(\gamma)} Q^{-1}(\varepsilon), \tag{A.4}$$

where (a) is a Gaussian approximation introduced in [75] that holds tight for $n \geq 100$ channel uses. Herein,

$$C(\gamma) = \log_2(1 + \gamma) \tag{A.5}$$

is the Shannon capacity, and

$$V(\gamma) = 1 - \frac{1}{(1 + \gamma)^2} \log_2^2 e \tag{A.6}$$

is the channel dispersion, which measures the stochastic variability of the channel relative to a deterministic channel with the same capacity. In addition, $Q^{-1}(\cdot)$ denotes the inverse Q-function.

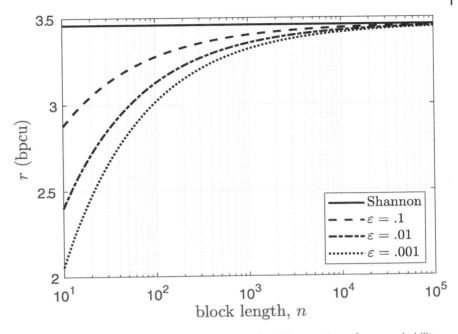

Figure A.1 Rate r as function of the block length n for different values of error probability ε and $\gamma = 10$ dB.

Figure A.1 illustrates the results from (A.4) assuming an AWGN channel with SNR of 10 dB. Notice that as the block length grows, $n \to \infty$, the achievable rate r approaches the Shannon capacity as expected. Also notice that for small block length, for instance $n < 1000$, there is a large performance gap between finite and infinite case, and this gap grows larger as ε increases.

Now, we can rewrite (A.4) as

$$\varepsilon(\gamma, b, n) \approx Q\left(\frac{C(\gamma) - b/n}{\sqrt{V(\gamma)/n}}\right), \tag{A.7}$$

which is the maximum error probability when transmitting b information bits over a channel with SNR γ and using n complex symbols. Notice that (A.7) matches the asymptotic *outage probability* when $n \to \infty$ and/or $\gamma \to 0$. Meanwhile, for block fading channels, the average maximum error probability is [292]

$$\bar{\varepsilon}(b, n) \approx \mathbb{E}_\gamma\left[Q\left(\frac{C(\gamma) - b/n}{\sqrt{V(\gamma)/n}}\right)\right] \tag{A.8}$$

since the channel becomes conditionally Gaussian on γ, and we only require to take expectation over the SNR to attain the corresponding average error probability. However, the effect of the fading on (A.8) induces a vanishing impact of the finite block length

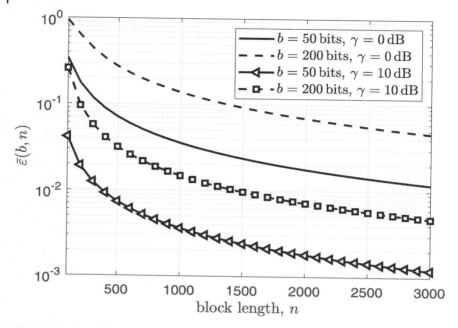

Figure A.2 Average maximum error probability as function of the block length n for different SNRs and payload.

under two cases [236]: 1. when b is not extremely small, and 2. there is not a strong LOS component (e.g., as in Rayleigh fading). Thus, the asymptotic outage probability, which is the Laplace approximation of (A.8), is a good match in such cases and is given by

$$\bar{\varepsilon}(b, n) \approx \mathbb{P}\left[\gamma < 2^{b/n} - 1\right] = F_\gamma(2^{b/n} - 1). \tag{A.9}$$

The behavior of (A.8) is illustrated in Figure A.2, which shows the average maximum error probability, $\bar{\varepsilon}(b, n)$, (often called outage probability) as function of the block length n. Note that increases in the block length n or increases in the SNR lead to a decrease in the outage probability. However, increases in the message payload b worsen the outage probability because they induce a higher rate since $r = \frac{b}{n}$.

Appendix B: Distribution of Transferred RF Energy Under CSI-free WET

B.1 Proof of Theorem 4.2

Departing from (4.28) and using (4.24), the RF power available as input to the energy harvester obeys

$$
\begin{aligned}
E_{\text{aa-ss}} &= \frac{\varrho}{M} \left| \mathbf{1}^T \mathbf{h}_x + \mathrm{i} \mathbf{1}^T \mathbf{h}_y \right|^2 \\
&= \frac{\varrho}{M} \left[\left(\mathbf{1}^T \mathbf{h}_x \right)^2 + \left(\mathbf{1}^T \mathbf{h}_y \right)^2 \right] \\
&\overset{(a)}{=} \frac{\varrho R_\Sigma}{2(\kappa + 1)M} \left(q_x^2 + q_y^2 \right) \\
&\overset{(b)}{\sim} \frac{\varrho R_\Sigma}{2(\kappa + 1)M} \chi^2 \left(2, \frac{2\kappa \left(v_1(\boldsymbol{\psi}, \theta)^2 + v_2(\boldsymbol{\psi}, \theta)^2 \right)}{R_\Sigma} \right),
\end{aligned}
\tag{B.1}
$$

where $R_\Sigma = \mathbf{1}^T \mathbf{R} \, \mathbf{1}$, while v_1 and v_2 are given in (4.31) and (4.32), respectively. Result in (a) comes from setting

$$
q_{x,y} = \mathbf{1}^T \mathbf{h}_{x,y} \sqrt{\frac{2(\kappa + 1)}{R_\Sigma}} \sim \mathcal{N}\left(\sqrt{\frac{\kappa}{R_\Sigma}} (v_1 \pm v_2), 1 \right).
\tag{B.2}
$$

Note that $\mathbf{1}^T \mathbf{h}_{x,y}$ is a Gaussian random variable, where the mean is equal to the sum of the means of each element of vector $\mathbf{h}_{x,y}$, while the variance equals the sum of the elements of the covariance matrix $\frac{1}{2(\kappa+1)} \mathbf{R}$ [294]. Therefore, the mean and variances are $\sqrt{\frac{\kappa}{2(\kappa+1)}} (v_1 \mp v_2)$ and $\frac{R_\Sigma}{2(\kappa+1)}$, respectively. Then, (b) comes from using the definition of a non-central chi-squared distribution [224, Cap.2] along with simple algebraic transformations. Finally, after using (4.30) we attain (4.29).

Wireless RF Energy Transfer in the Massive IoT Era: Towards Sustainable Zero-energy Networks. First Edition. Onel Alcaraz López and Hirley Alves.
© 2022 by The Institute of Electrical and Electronics Engineers, Inc. Published 2022 by John Wiley & Sons, Inc. Companion website: www.wiley.com/go/Alves/WirelessEnergyTransfer

B.2 Proof of Theorem 4.4

Based on (4.44), the RF energy available at S under the AA−IS operation is given by

$$E_{\text{aa}-\text{is}} = \frac{\varrho}{M}\left(\mathbf{h}_x^T\mathbf{h}_x + \mathbf{h}_y^T\mathbf{h}_y\right), \tag{B.3}$$

for which we analyze its distribution as follows. Let us define

$$\mathbf{z}_{x,y} = \sqrt{2(\kappa + 1)}\mathbf{R}^{-1/2}\left(\mathbf{h}_{x,y} - \sqrt{\frac{\kappa}{2(\kappa + 1)}}\boldsymbol{\omega}_{x,y}\right), \tag{B.4}$$

which is distributed as $\mathcal{N}(0, \mathbf{I})$, then

$$\mathbf{h}_{x,y} = \frac{1}{\sqrt{2(\kappa + 1)}}\mathbf{R}^{1/2}\mathbf{z}_{x,y} + \sqrt{\frac{\kappa}{2(\kappa + 1)}}\boldsymbol{\omega}_{x,y}, \tag{B.5}$$

$$\mathbf{h}_{x,y}^T\mathbf{h}_{x,y} = \frac{1}{2(\kappa + 1)}\left(\mathbf{R}^{1/2}\mathbf{z}_{x,y} + \sqrt{\kappa}\boldsymbol{\omega}_{x,y}\right)^T\left(\mathbf{R}^{1/2}\mathbf{z}_{x,y} + \sqrt{\kappa}\boldsymbol{\omega}_{x,y}\right)$$

$$= \frac{1}{2(\kappa + 1)}\left(\mathbf{z}_{x,y} + \mathbf{R}^{-1/2}\sqrt{\kappa}\boldsymbol{\omega}_{x,y}\right)^T\mathbf{R}\left(\mathbf{z}_{x,y} + \mathbf{R}^{-1/2}\sqrt{\kappa}\boldsymbol{\omega}_{x,y}\right), \tag{B.6}$$

where last step comes from simple algebraic transformations. Notice that $\mathbf{R} = \mathbf{Q}\boldsymbol{\Lambda}\mathbf{Q}^T$ is the spectral decomposition of $\mathbf{R}$, where $\boldsymbol{\Lambda}$ is a diagonal matrix containing the eigenvalues of $\mathbf{R}$, and $\mathbf{Q}$ is a matrix whose column vectors are the orthogonalized eigenvectors of $\mathbf{R}$. In order to find the eigenvalues, $\{\lambda_j\}$, of $\mathbf{R}$, we require solving $\det(\mathbf{R} - \lambda\mathbf{I}) = 0$ for λ, which is analytically intractable for a general matrix $\mathbf{R}$. However, there is analytical tractability for the special case of uniform spatial correlation. Specifically, in case the antenna elements are correlated between each other with coefficient ρ, i.e., $R_{i,j} = \rho$, $\forall i \neq j$, we have that

$$R_{\Sigma} = M\left(1 + (M - 1)\rho\right). \tag{B.7}$$

Also, in order to guarantee that $\mathbf{R}$ is positive semi-definite and consequently a viable covariance matrix, ρ is lower bounded by $-\frac{1}{M-1}$ [295], thus $-\frac{1}{M-1} \leq \rho \leq 1$ and $0 \leq R_{\Sigma} \leq M^2$. Then, under such a correlation model we have that [84, Eq.(32)]

$$\boldsymbol{\Lambda} = \text{diag}\left(1 - \rho, \cdots, 1 - \rho, 1 + (M - 1)\rho\right), \tag{B.8}$$

thus, matrix $\mathbf{R}$ is characterized by two different eigenvalues: $1 - \rho$, which is independent of the number of antennas but has multiplicity $M - 1$, and $1 + (M - 1)\rho$, whose multiplicity is 1 but increases linearly with M. Then, matrix $\mathbf{Q}^T$ can be written as

$$\mathbf{Q}^T = \begin{bmatrix} -\dfrac{1}{\sqrt{1\times 2}} & 0 & 0 & \cdots & 0 & \dfrac{1}{\sqrt{1\times 2}} \\[2mm] -\dfrac{1}{\sqrt{2\times 3}} & 0 & 0 & \cdots & \dfrac{2}{\sqrt{2\times 3}} & -\dfrac{1}{\sqrt{2\times 3}} \\[2mm] -\dfrac{1}{\sqrt{3\times 4}} & 0 & 0 & \cdots & -\dfrac{1}{\sqrt{3\times 4}} & -\dfrac{1}{\sqrt{3\times 4}} \\[2mm] \vdots & \vdots & \vdots & \ddots & \vdots & \vdots \\[2mm] -\dfrac{1}{\sqrt{(M-1)M}} & \dfrac{M-1}{\sqrt{(M-1)M}} & -\dfrac{1}{\sqrt{(M-1)M}} & \cdots & -\dfrac{1}{\sqrt{(M-1)M}} & -\dfrac{1}{\sqrt{(M-1)M}} \\[2mm] \dfrac{1}{\sqrt{M}} & \dfrac{1}{\sqrt{M}} & \dfrac{1}{\sqrt{M}} & \cdots & \dfrac{1}{\sqrt{M}} & \dfrac{1}{\sqrt{M}} \end{bmatrix}.$$

$$(B.9)$$

Now, by substituting $\mathbf{R} = \mathbf{Q}\boldsymbol{\Lambda}\mathbf{Q}^T$ into (B.6), while setting $\zeta_{x,y} = 2(\kappa + 1)\mathbf{h}_{x,y}^T\mathbf{h}_{x,y}$, we obtain

$$
\begin{aligned}
\zeta_{x,y} &= \left(\mathbf{z}_{x,y} + \sqrt{\kappa}\left(\mathbf{Q}\boldsymbol{\Lambda}\mathbf{Q}^T\right)^{-\frac{1}{2}}\boldsymbol{\omega}_{x,y}\right)^T \mathbf{Q}\boldsymbol{\Lambda}\mathbf{Q}^T\left(\mathbf{z}_{x,y} + \sqrt{\kappa}\left(\mathbf{Q}\boldsymbol{\Lambda}\mathbf{Q}^T\right)^{-\frac{1}{2}}\boldsymbol{\omega}_{x,y}\right) \\
&\overset{(a)}{=} \left(\mathbf{Q}^T\mathbf{z}_{x,y} + \sqrt{\kappa}\boldsymbol{\Lambda}^{-\frac{1}{2}}\mathbf{Q}^T\boldsymbol{\omega}_{x,y}\right)^T \boldsymbol{\Lambda}\left(\mathbf{Q}^T\mathbf{z}_{x,y} + \sqrt{\kappa}\boldsymbol{\Lambda}^{-\frac{1}{2}}\mathbf{Q}^T\boldsymbol{\omega}_{x,y}\right) \\
&\overset{(b)}{=} \left(\mathbf{P}_{x,y} + \sqrt{\kappa}\mathbf{u}_{x,y}\right)^T \boldsymbol{\Lambda}\left(\mathbf{P}_{x,y} + \sqrt{\kappa}\mathbf{u}_{x,y}\right) \\
&\overset{(c)}{=} \sum_{j=1}^M \lambda_j\left(c_j + \sqrt{\kappa}u_{x,y_j}\right)^2,
\end{aligned}
\tag{B.10}
$$

where (a) comes after some algebraic transformations, (b) follows from taking $\mathbf{P}_{x,y} = \mathbf{Q}^T\mathbf{z}_{x,y} \sim \mathcal{N}(0, \mathbf{I})$ and

$$
\mathbf{u}_{x,y} = \boldsymbol{\Lambda}^{-\frac{1}{2}}\mathbf{Q}^T\boldsymbol{\omega}_{x,y} = \boldsymbol{\Lambda}^{-\frac{1}{2}}\begin{bmatrix} \frac{1}{\sqrt{2}}(\omega_M - \omega_1) \\[1mm] \frac{1}{\sqrt{2\times 3}}(2\omega_{M-1} - \omega_1 - \omega_M) \\[1mm] \frac{1}{\sqrt{3\times 4}}\left(3\omega_{M-2} - \omega_1 - \sum_{j=M-1}^M \omega_j\right) \\[1mm] \frac{1}{\sqrt{4\times 5}}\left(4\omega_{M-3} - \omega_1 - \sum_{j=M-2}^M \omega_j\right) \\[1mm] \vdots \\[1mm] \frac{1}{\sqrt{(M-2)(M-1)}}\left((M-2)\omega_3 - \omega_1 - \sum_{j=4}^M \omega_j\right) \\[1mm] \frac{1}{\sqrt{(M-1)M}}\left((M-1)\omega_2 - \omega_1 - \sum_{j=3}^M \omega_j\right) \\[1mm] \frac{1}{\sqrt{M}}\sum_{j=1}^M \omega_j \end{bmatrix},
\tag{B.11}
$$

$$
u_{x,y_j} = \begin{cases} \dfrac{j\omega_{M-j+1} - \omega_1 - \sum_{t=M-j+2}^M \omega_t}{\sqrt{j(j+1)\lambda_j}}, & \text{for } j \leq M-1 \\[3mm] \dfrac{1}{\sqrt{M\lambda_M}}\sum_{t=1}^M \omega_t, & \text{for } j = M \end{cases},
\tag{B.12}
$$

which come from using (B.8) and (B.9); while (c) follows after taking $c_j \sim \mathcal{N}(0, 1)$.

By incorporating these results into (B.3) we obtain the distribution of E_{aa-is} as follows

$$E_{aa-is} \sim \frac{\varrho}{2M(\kappa+1)}(\zeta_x + \zeta_y)$$

$$\sim \frac{\varrho}{2M(\kappa+1)} \sum_{j=1}^{M} \lambda_j \left(\left(c_j + \sqrt{\kappa}u_{x_j}\right)^2 + \left(\tilde{c}_j + \sqrt{\kappa}u_{y_j}\right)^2 \right)$$

$$\sim \frac{\varrho}{2M(\kappa+1)} \left((1-\rho)\chi^2 \left(2(M-1), \frac{2\kappa\tilde{v}(\boldsymbol{\psi},\theta)}{1-\rho} \right) + \right.$$

$$\left. + \left(1 + (M-1)\rho\right)\chi^2 \left(2, \frac{\kappa\beta(\boldsymbol{\psi},\theta)}{M\big(1+(M-1)\rho\big)} \right) \right) \tag{B.13}$$

since $\tilde{c}_j \sim \mathcal{N}(0,1)$,

$$u_{xM}^2 + u_{yM}^2 = \frac{1}{M\lambda_M} \left(\left(\sum_{t=1}^{M} \omega_{xt} \right)^2 + \left(\sum_{t=1}^{M} \omega_{yt} \right)^2 \right)$$

$$= \frac{\left(v_1(\boldsymbol{\psi},\theta) - v_2(\boldsymbol{\psi},\theta)\right)^2 + \left(v_1(\boldsymbol{\psi},\theta) + v_2(\boldsymbol{\psi},\theta)\right)^2}{M\big(1+(M-1)\rho\big)}$$

$$= \frac{2\beta(\boldsymbol{\psi},\theta)}{M\big(1+(M-1)\rho\big)}, \tag{B.14}$$

while $\tilde{v} = \frac{1}{2}\sum_{j=1}^{M-1} \lambda_j \left(u_{x_j}^2 + u_{y_j}^2\right)$, which appears expanded as a function of $\boldsymbol{\psi}$ and $\boldsymbol{\phi}$ in (B.15) and it is obtained by (a) using (B.12), (b) expanding the quadratic binomials, and (c) performing some algebraic simplifications by taking advantage of $\sin^2 a + \cos^2 a = 1$. Then, we attain (4.46) by regrouping terms and performing further algebraic simplifications by taking advantage of $\cos a \cos b + \sin a \sin b = \cos(a-b)$.

Notice that (B.13) holds under the assumption of uniform spatial correlation. However, given that for the AA–SS scheme we were able of writing the distribution merely as a function of parameter R_Σ, which is not linked to any specific kind of correlation, we can expect that the behavior under the AA–IS scheme depends, at least approximately, on R_Σ rather on the specific entries of matrix $\mathbf{R}$. In fact, such a hypothesis has been shown to be accurate under the scenario of LOS components with equal mean phases studied in [84]. Then,

$$\tilde{v}(\boldsymbol{\psi},\theta) \overset{(a)}{=} \sum_{j=1}^{M-1} \frac{1}{2j(j+1)} \left(\left(j\omega_{xM-j+1} - \omega_{x1} - \sum_{t=M-j+2}^{M} \omega_{xt} \right)^2 + \left(j\omega_{yM-j+1} - \omega_{y1} - \sum_{t=M-j+2}^{M} \omega_{yt} \right)^2 \right)$$

$$\overset{(b)}{=} \sum_{j=1}^{M-1} \frac{1}{2j(j+1)} \left(j^2 \left(\omega_{xM-j+1}^2 + \omega_{yM-j+1}^2 \right) + 2 + \left(\sum_{t=M-j+2}^{M} \omega_{xt} \right)^2 + \left(\sum_{t=M-j+2}^{M} \omega_{yt} \right)^2 + \right.$$

$$-2j\left(\omega_{xM-j+1}+\omega_{yM-j+1}\right)+2\sum_{t=M-j+2}^{M}\left(\omega_{xt}+\omega_{yt}\right)+$$

$$-2j\left(\omega_{xM-j+1}\sum_{t=M-j+2}^{M}\omega_{xt}+\omega_{yM-j+1}\sum_{t=M-j+2}^{M}\omega_{yt}\right)\Bigg)$$

$$\overset{(c)}{=}\sum_{j=1}^{M-1}\frac{1}{j(j+1)}\left(\left(\sum_{t=M-j+1}^{M-1}\cos\left(\psi_t+\phi_t\right)\right)^2+\left(\sum_{t=M-j+1}^{M-1}\sin\left(\psi_t+\phi_t\right)\right)^2+\right.$$

$$+j^2+1-2j\cos\left(\psi_{M-j}+\phi_{M-j}\right)-2j\Bigg(\cos\left(\psi_{M-j}+\phi_{M-j}\right)\sum_{t=M-j+1}^{M-1}\cos\left(\psi_t+\phi_t\right)+$$

$$\left.+\sin\left(\psi_{M-j}+\phi_{M-j}\right)\sum_{t=M-j+1}^{M-1}\sin\left(\psi_t+\phi_t\right)\Bigg)+2\sum_{t=M-j+1}^{M-1}\cos\left(\psi_t+\phi_t\right)\right). \tag{B.15}$$

---

we substitute $\rho = \frac{R_\Sigma - M}{M(M-1)}$ coming from (B.7) into (B.13) such that we attain (4.45).

Validation. To evaluate the accuracy of (4.45) we utilize the Bhattacharyya distance metric [296], which measures the similarity of two probability distributions $\mathbf{f}_1$ and $\mathbf{f}_2$, and it is given by

$$d_B(\mathbf{f}_1, \mathbf{f}_2) = -\ln\left(c_B(\mathbf{f}_1, \mathbf{f}_2)\right), \tag{B.16}$$

where c_B is the Bhattacharyya coefficient [296]. In the case of our interest, both probability distributions characterize E_{aa-is}. However, for $\mathbf{f}_1$ we assume a uniform spatial correlation matrix $\mathbf{R}$, thus we may use (4.45) which is exact in such scenario; while for $\mathbf{f}_2$ we assume a randomly generated correlation matrix $\mathbf{R}'$ such that $R_\Sigma = R'_\Sigma$, thus, we evaluate the distribution directly from (B.3). We utilize the histogram formulation for estimating $\mathbf{f}_1$ and $\mathbf{f}_2$. Specifically, we estimate $\hat{\mathbf{f}}_1 = \{\hat{f}_{1,i}\}_{i=1,\cdots,m}$ (with $\sum_{i=1}^{m}\hat{f}_{1,i} = 1$) and $\hat{\mathbf{f}}_2 = \{\hat{f}_{2,i}\}_{i=1,\cdots,m}$ (with $\sum_{i=1}^{m}\hat{f}_{2,i} = 1$), where m is the number of histogram bins. Then, the Bhattacharyya coefficient is calculated as [296]

$$c_B(\hat{\mathbf{f}}_1, \hat{\mathbf{f}}_2) = \sum_{i=1}^{m}\sqrt{\hat{f}_{1,i}\hat{f}_{2,i}}. \tag{B.17}$$

By substituting (B.17) into (B.16) we calculate the similarity between distributions. Note that according to (4.45) we have that $d_B(\mathbf{f}_1, \mathbf{f}_2) \in [0, \infty)$, where 0 corresponds to the case when $\mathbf{f}_1 = \mathbf{f}_2$.

Figure B.1(a) shows the average Bhattacharyya distance, $\bar{d}_B$, as a function of θ for $\kappa = 10, M \in \{4, 8\}, \boldsymbol{\psi} = \mathbf{0}$ (no preventive phase shifting) and $\boldsymbol{\psi}$ taken uniformly random from $[0, 2\pi]^{M-1}$ to account for different possible preventive phase shiftings. We generated 1000 random correlation matrices and averaged over the Bhattacharyya distance for the distributions corresponding to each of them. We utilized 2×10^5 distribution

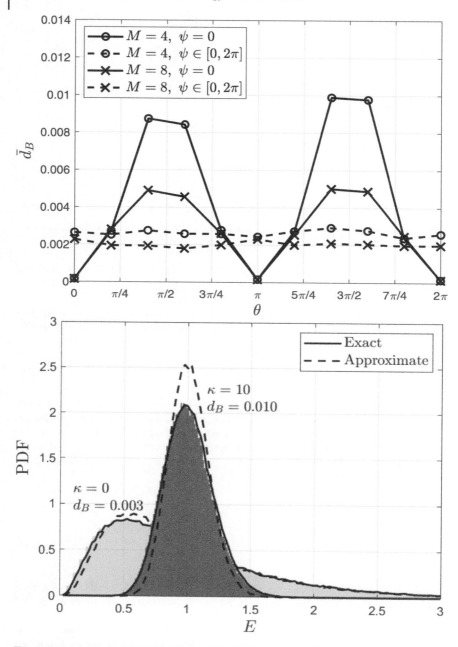

Figure B.1 (a) Average Bhattacharyya distance between $\hat{f}_1$ and $\hat{f}_2$ as a function of θ for $\kappa = 10$, $M \in \{4,8\}$, $\boldsymbol{\psi} = 0$ (no preventive phase shifting), and $\boldsymbol{\psi}$ taken uniformly random from $[0,2\pi]^{M-1}$ (top). (b) Monte Carlo-based comparison between the exact f_2 and proposed approximate f_1 distributions of the incident RF power under the AA–IS scheme. We set $M = 4$, $\theta = 3\pi/2$ and $\boldsymbol{\psi} = \mathbf{0}$. The comparison is carried out for two different cases: 1. $\kappa = 0$, $\bar{d}_B = 0.003$; and 2. $\kappa = 10$, $\bar{d}_B = 0.010$ (bottom).

samples and set $m = 240$, while the histogram edges were uniformly chosen between 0 and 6, which is an appropriate range since, without loss of generality, we used $\varrho = 1$. Although we present the results for a specific value of κ ($\kappa = 10$), the trends remain similar and the specific Bhattacharyya distance values are often smaller for smaller κ, while no significant deviations are experienced for larger κ. As observed in Figure B.1(a), the largest difference between both distributions is when $\boldsymbol{\psi} = \mathbf{0}$, M is small, e.g., $M = 4$, and $\theta \approx \pi/4 \pm \pi/2$. In such a scenario, d_B values may become close to 0.012 but they never surpass such a limit. Meanwhile, for other parameter configurations $\bar{d}_B$ becomes significantly smaller.

We select $\bar{d}_B = 0.010$ and $\bar{d}_B = 0.002$ to illustrate the cases of respectively large and small differences between the considered distributions in Figure B.1(b). Notice that the approximate distributions indeed approach the exact ones, even with greater accuracy for smaller d_B as expected. Finally, since $\bar{d}_B = 0.010$ is a very pessimistic assumption as it is only reachable for few values of θ, results in Figure B.1(b) validate the accuracy of (4.45) in general scenarios.

Appendix C: Clustering Algorithms

The process of grouping similar data points in an unsupervised way is called *clustering*. Clustering algorithms use a certain distance or dissimilarity metric in order to separate observations into different groups. The concept of distance or dissimilarity is the essential component of any form of clustering [297]. Formally, the distance $d(c_i, c_j)$ between c_i and c_j is considered to be a two argument function satisfying the following conditions:

1. $d(c_i, c_j) \geq 0, \forall c_i, c_j$;
2. $d(c_i, c_i) = 0, \forall c_i$;
3. $d(c_i, c_j) = d(c_j, c_i), \forall c_i, c_j$;
4. $d(c_i, c_j) + d(c_j, c_z) \geq d(c_i, c_z), \forall c_i, c_j, c_z$.

For example:

- *Euclidean distance*, which corresponds to the length of a line segment between the two data points: $d(c_i, c_j) = ||c_i - c_j||$;
- *Manhattan distance*, which corresponds to the L1 norm distance between the two data points: $d(c_i, c_j) = ||c_i - c_j||_1$;
- *Cosine coefficient* relates the overlap to the geometric average of the two sets: $d(c_i, c_j) = 1 - \frac{c_i^T c_j}{||c_i||||c_j||}$.

On the other hand, a dissimilarity function satisfies $1 - 3$ but not necessarily 4. The most common methods of clustering are discussed next.

C.1 Partitioning Methods

These methods involve partitioning the data while grouping similar items. Common partitioning algorithms are K-Means, K-Medoids, and K-Modes. Results vary according to the pre-defined number of clusters that are allowed. However, when such a value is uncertain and/or can be flexibly selected, the Elbow method or Silhouette index [298] can assist in the corresponding optimization.

Wireless RF Energy Transfer in the Massive IoT Era: Towards Sustainable Zero-energy Networks.
First Edition. Onel Alcaraz López and Hirley Alves.
© 2022 by The Institute of Electrical and Electronics Engineers, Inc. Published 2022 by John Wiley & Sons, Inc. Companion website: www.wiley.com/go/Alves/WirelessEnergyTransfer

C.1.1 *K*-Means

K-Means is maybe the most widely known clustering approach. It iteratively relocates the cluster centers by computing the mean position of a cluster as follows:

- Initially, k cluster centers are randomly set.
- The distance between each data point and cluster centers is calculated, and each data point is assigned to its closest (according to the adopted distance measure) cluster head/center.
- Clusters centers are re-computed. The computation depends on the distance measure being used, e.g., the centers correspond to the mean of the associated clusters' points in case of Euclidean distance, the component-wise median in case of Manhattan distance, and the normalized mean in case of Cosine coefficient.
- This process is repeated until the cluster centers do not change.

Intuitively, since path loss, thus, receive power, depends strictly on the Euclidean distance, this latter seems the most suitable measure to address the PBs deployment clustering problem discussed in Chapter 5. Indeed, the interested reader can corroborate via simulations that *K*-Means clustering with Euclidean distance outperforms the variants with Manhattan distance and Cosine coefficient for typical PBs deployment problems. However, clusters' centroids may not be necessarily set as the mean of the associated points. In fact, in Chapter 5, we show an alternative computation method for the cluster centers that is suitable for a fair PBs deployment strategy.

It is worth mentioning that the *K*-Means method performs extremely well on huge data sets, i.e., it is scalable. However, it is very sensitive to outliers, which have a severe impact while computing the means of the cluster data points. Also, results vary according to the initial choice of cluster centers. Finally, note that the *K*-Means algorithm works well only for spherical data and fails to perform well on arbitrary shapes of data.

C.1.2 *K*-Medoids

K-Medoids is similar to *K*-Means. Both attempt to minimize a kind of *separation* between points labeled to be in a cluster and a point designated as the center of that cluster. In contrast to the *K*-Means algorithm, *K*-Medoids chooses actual data points as centers (medoids or exemplars), and can be used with arbitrary dissimilarity measures (while *K*-Means uses only distance measures). Because of the latter, *K*-Medoids is more robust to noise and outliers than *K*-Means. The medoid of a cluster is defined as the object in the cluster whose average dissimilarity to all the objects in the cluster is minimal, i.e., the most centrally located point in the cluster.

C.1.3 *K*-Modes

K-Modes clustering is also similar to *K*-Means but for categorical data, thus, instead of calculating the mean of the cluster data points, it calculates the mode (the value that occurs most frequently) of the cluster.

C.2   Hierarchical Methods

These methods involve decomposing the data hierarchically, and differently from partitioning methods, the number of clusters is not specified in advance. Hierarchical clustering helps to determine the optimal number of clusters. There are two main types of hierarchical methods [299]:

- Agglomerative (*bottom-up* approach), where each data point starts in its own cluster, and pairs of clusters are merged as one moves up the hierarchy.
- Divisive (*top-down* approach), where all data points start in one cluster, and splits are performed recursively as one moves down the hierarchy.

In general, the merges and splits are determined in a greedy manner, and the results are usually presented in a dendrogram.

In order to decide which clusters should be combined (for agglomerative), or where a cluster should be split (for divisive), a measure of dissimilarity between sets of observations is required. This is often achieved by using an appropriate distance metric (a measure of distance between pairs of observations), and a linkage criterion which specifies the dissimilarity of sets as a function of the pairwise distances of observations in the sets. Let us see, for instance, how it works in the case of agglomerative hierarchical clustering:

- Each data point constitutes its own cluster, i.e., singleton cluster.
- The two clusters that are closest to each other are merged into a cluster. This requires computing distances between data points (as exemplified at the beginning of this appendix), and also linkage criteria, e.g., single/complete/average linkage (minimum/maximum/average distance between clusters), centroid linkage (distance between the centroids of both clusters), Ward's criterion (to minimize the total within-cluster variance and find the pair of clusters that leads to minimum increase in total within-cluster variance after merging). The consequence is that there is one less cluster.
- The distances between the new and old clusters are recalculated/updated.
- The last two steps are repeated until all clusters are merged into one single cluster including all data points.

C.3   Other Methods

Other clustering methods are for instance:

- Density-based methods, which are used in anomaly detection. The data points that are highly dense are grouped together, thus leaving the data points with low density. Examples of these methods are DBSCAN (Density-Based Spatial Clustering of Application with Noise) and OPTICS (Ordering Points To Identify Cluster Structures). The former has the ability to perform well on data with arbitrary shapes, while the latter is similar to DBSCAN but surmounts its difficult-to-tune hyperparameters disadvantage.
- Model-based methods, which involve applying a model to find the best cluster structures, e.g., by using an expectation-maximization algorithm.

However, the application of these methods in the PBs deployment optimization is, at least, not evident, and we do not delve further into them.

C.4 Pre-processing

Pre-processing techniques such as normalization, scaling, and thinning are often required before using clustering algorithms [300], especially when the components of the data point vectors do not correspond to the same physical units and/or extreme outliers appear in large data sets. However, none of these hold for the clustering applications discussed in this book, and such pre-processing is not required.

Transformation (other than normalization or standardization for heterogeneous data sets) constitutes another kind of pre-processing technique. As an example, consider the problem of clustering angular data as required in Section 4.3.3. Let us say we have the following data $\{-3\pi/4, -\pi/4, 0, 5\pi/4, \pi\}$, e.g., devices' angular position with respect to a PB, and we need to form two clusters. If distance-based clustering is applied directly, one may get the erroneous clustering: $\{-3\pi/4, -\pi/4, 0\}$, $\{5\pi/4, \pi\}$. However, note than $-3\pi/4$ and π are in fact closer than $-3\pi/4$ and $-\pi/4$ and they should be clustered together. A nice and simple approach is to take each angle and use it to create a point on the unit circle in two dimensions, i.e., given the angle a_i, create the data point $c_i = (\cos a_i, \sin a_i)$, and then cluster all the coordinates $\{c_i\}$. This results in the following, more practical, clustering: $\{-3\pi/4, 5\pi/4, \pi\}$, $\{-\pi/4, 0\}$.

Appendix D: Required SNR for a Target Decoding Error Probability (Proof of Theorem 6.1)

First, observe that (6.14) is obtained by properly re-writing (A.4) while using $\mathcal{V}(x) = \sqrt{V(x)}\ln(2)$. Then, (6.14) could be stated as $f(\gamma_{th}) = k(\gamma_{th}) - \gamma_{th} = 0$, where $k(\gamma_{th}) = q_1 q_2^{\mathcal{V}(\gamma_{th})} - 1$, $q_1 = 2^{b/n} \geq 1$ since $r = b/n \geq 0$, and $q_2 = \exp\left(\frac{1}{\sqrt{n}}Q^{-1}(\varepsilon_0)\right)$. Now, note that $f(\gamma_{th})$ is continuous, while $f(0) = q_1 - 1 \geq 0$ and $\lim_{\gamma_{th}\to\infty} f(\gamma_{th}) = -\infty$, thus, there is at least one γ_{th} such that $f(\gamma_{th}) = 0$. Based on the equation to solve, e.g. $k(\gamma_{th}) = \gamma_{th}$ with $\gamma_{th} \in \mathbb{R}^+$, observe that for $\varepsilon_0 < 0.5$ we have $Q^{-1}(\varepsilon_0) > 0$, thus both γ_{th} and $k(\gamma_{th})$ are increasing functions. Taking the derivatives of $k(\gamma_{th})$ yields

$$\frac{d}{d\gamma_{th}}k(\gamma_{th}) = \frac{q_1 q_2^{\mathcal{V}(\gamma_{th})}\ln(q_2)}{(1+\gamma_{th})^3 \mathcal{V}(\gamma_{th})}, \tag{D.1}$$

$$\frac{d^2}{d\gamma_{th}^2}k(\gamma_{th}) = -\frac{\ln(q_2)q_1 b^{\mathcal{V}(\gamma_{th})}}{\mathcal{V}(\gamma_{th})(1+\gamma_{th})^2}\left[\frac{1}{(1+\gamma_{th})^2\mathcal{V}(\gamma_{th})} + 3 - \frac{\ln(q_2)}{(1+\gamma_{th})^2}\right], \tag{D.2}$$

where $\ln(q_2) > 0$. Thus, we can claim that $k(\gamma_{th})$ is concave if

$$\frac{1}{(1+\gamma_{th})^2\mathcal{V}(\gamma_{th})} + 3 > \frac{\ln(q_2)}{(1+\gamma_{th})^2}$$

$$q_2 < \exp\left(\frac{1}{\mathcal{V}(\gamma_{th})} + 3(1+\gamma_{th})^2\right)$$

$$Q^{-1}(\varepsilon_0) < \sqrt{n}\left(\frac{1}{\mathcal{V}(\gamma_{th})} + 3(1+\gamma_{th})^2\right)$$

$$\varepsilon_0 > Q\left(\sqrt{n}\left(\frac{1}{\mathcal{V}(\gamma_{th})} + 3(1+\gamma_{th})^2\right)\right), \tag{D.3}$$

where the right side is maximized for the minimum value of $\sqrt{n}\left(\frac{1}{\mathcal{V}(\gamma_{th})} + 3(1+\gamma_{th})^2\right)$. Setting $n = 100$ channel uses, which is the minimum value for which the finite block length error probability approximation in (A.4) is valid, and $\gamma_{th} = 0.1655$, which minimizes the remaining terms, we reach $\varepsilon_0 > Q(46.6364) \approx 4.4 \times 10^{-475}$. Evidently, that requirement is met for any setup of practical interest. Thus, $k(\gamma_{th})$ is increasing and concave, and since $k(0) > 0$, which is the starting point of line γ_{th}, we conclude that they intersect at one point only. Therefore, the solution is unique.

Wireless RF Energy Transfer in the Massive IoT Era: Towards Sustainable Zero-energy Networks. First Edition. Onel Alcaraz López and Hirley Alves. © 2022 by The Institute of Electrical and Electronics Engineers, Inc. Published 2022 by John Wiley & Sons, Inc. Companion website: www.wiley.com/go/Alves/WirelessEnergyTransfer

Thus, we can say that the unique solution of (6.14) is a fixed point of such an equation. Based on the fixed point theory [301], if $|k(\gamma_{\text{th}})| < 1$, the fixed point iteration in (6.14) will converge to the solution. Using (D.1) and performing some algebraic transformations, yields

$$
\left| \frac{d}{d\gamma_{\text{th}}} k(\gamma_{\text{th}}) \right| = \left| \frac{q_1 q_2^{\mathcal{V}(\gamma_{\text{th}})} \ln(q_2)}{(1 + \gamma_{\text{th}})^3 \mathcal{V}(\gamma_{\text{th}})} \right|
$$

$$
\stackrel{(a)}{=} \frac{2^r \exp\left(\frac{\mathcal{V}(\gamma_{\text{th}})}{\sqrt{n}} Q^{-1}(\varepsilon_0) \right) \frac{|Q^{-1}(\varepsilon_0)|}{\sqrt{n}}}{(1 + \gamma_{\text{th}})^3 \mathcal{V}(\gamma_{\text{th}})}
$$

$$
\stackrel{(b)}{=} \frac{2^r \exp\left(\frac{\log_2(1+\gamma_{\text{th}}) - r}{\log_2 e} \right) \frac{|\log_2(1+\gamma_{\text{th}}) - r|}{\mathcal{V}(\gamma_{\text{th}}) \log_2 e}}{(1 + \gamma_{\text{th}})^3 \mathcal{V}(\gamma_{\text{th}})}
$$

$$
\stackrel{(c)}{=} \frac{|\log_2(1 + \gamma_{\text{th}}) - r|}{\gamma_{\text{th}}(\gamma_{\text{th}} + 2) \log_2 e}, \tag{D.4}
$$

where *(a)* and *(b)* come from using the expressions of q_1 and q_2, and

$$
Q^{-1}(\varepsilon_0) = \frac{\log_2(1+\gamma_{\text{th}}) - r}{\frac{\mathcal{V}(\gamma_{\text{th}}) \log_2 e}{\sqrt{n}}}, \tag{D.5}
$$

respectively; while *(c)* is attained after substituting $\mathcal{V}(\gamma_{\text{th}}) = \sqrt{1 - \frac{1}{(1+\gamma_{\text{th}})^2}}$ followed by some simplifications. Notice that for $\varepsilon_0 \leq 0.5$, which is the case of practical interest, we have that $\log_2(1 + \gamma_{\text{th}}) \geq r$, thus

$$
\left| \frac{d}{d\gamma_{\text{th}}} k(\gamma_{\text{th}}) \right| < \frac{\log_2(1 + \gamma_{\text{th}})}{\gamma_{\text{th}}(\gamma_{\text{th}} + 2) \log_2 e} \leq \ln 2 < 1 \tag{D.6}
$$

since $\log_2(1 + \gamma_{\text{th}}) \leq \gamma_{\text{th}}(\gamma_{\text{th}} + 2)$ for $\gamma_{\text{th}} \geq 0$. Therefore, and from Banach's fixed point theorem [301], the (at least) linear convergence of a fixed-point iteration algorithm is guaranteed provided any initial point. In this particular case, we choose $\gamma_{\text{th}} = \infty$ ($\mathcal{V}(\infty) = 1$) as the initial point. □

D.1 On the Convergence of Algorithm 3

Figure D.1 shows the required number of iterations to solve Algorithm 3 as a function of the target error probability, ε_0, for setups with $n \in \{100, 500, 1000\}$ channel uses and $b = 32$ bytes. We can notice the very fast convergence of the iterative method in solving (6.14), even for a rigorous accuracy given by $\gamma_\Delta = 10^{-3}$, where γ_Δ constitutes the required SNR difference between consecutive iterations to terminate Algorithm 3 execution. As shown in the figure, small values of ε_0 require more iterations, especially for relatively large values of n. When n increases, the rate diminishes and the required γ_{th} becomes smaller, thus more iterations are necessary to solve the problem with the given accuracy. The convergence would be slower if we decrease γ_Δ. On the other hand, if we adopt a less demanding value of γ_Δ such as 10^{-2}, three iterations would be sufficient.

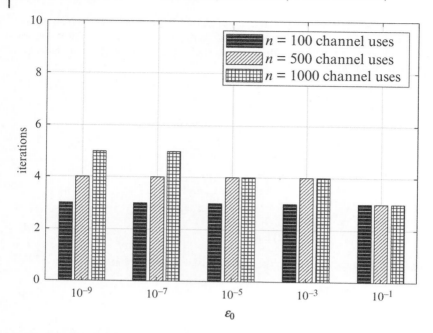

Figure D.1 Required number of iterations to solve Algorithm 3 as a function of ε_0 for $b = 32$ bytes, $n \in \{100, 500, 1000\}$ channel uses, and $\gamma_\Delta = 10^{-3}$.

One of the main advantages of Algorithm 3 is that it does not require an initial search interval, differently, for instance, from the bisection method. For certain search interval $I_{\gamma_{\text{th}}}$ on γ_{th}, and using the bisection method, we have that $\frac{I_{\gamma_{\text{th}}}}{2^{\text{iterations}+1}} \leq \gamma_\Delta$. Thus, the required number of iterations shall not be less than $\log_2\left(\frac{I_{\gamma_{\text{th}}}}{\gamma_\Delta}\right) - 1$. As an example, if we search on an interval of width $I_{\gamma_{\text{th}}} = 4$, for an accuracy of $\gamma_\Delta = 10^{-3}$ and $\gamma_\Delta = 10^{-2}$, 11 and 8 iterations would be respectively required under the bisection method, which are considerable higher than the required when using Algorithm 3. Finally, observe that the value of γ_{th} has to be updated only when some element in (n, b, ε_0) changes. Thus, it could be possible that Algorithm 3 does not run in S, but in another entity which broadcasts its value.

Bibliography

1 Ornes S. The internet of things and the explosion of interconnectivity. Proc Natl Acad Sci U S A. 2016;113(40):11059–11060.

2 Goasduff L. Gartner Says 5.8 Billion Enterprise and Automotive IoT Endpoints Will Be in Use in 2020; 2019. Available from: https://www.gartner.com/en/newsroom/press-releases/2019-08-29-gartner-says-5-8-billion-enterprise-and-automotive-io.

3 Grijpink F, Kutcher E, Ménard A, Ramaswamy S, Schiavotto D, Manyika J, et al. Connected world: An evolution in connectivity beyond the 5G revolution. McKinsey Glob Inst. 2020;(304):2.

4 Ericsson. Ericsson Technology Review: Spotlight on the Internet of Things; 2019.

5 Bosch Software Innovations. Industrial Internet: Putting the vision into practice. Bosch White Pap. 2015:1–14.

6 Nokia. Industrial IoT Networks: How 5G is transforming industry verticals. White Pap. 2020.

7 Frost and Sullivan. Future of Consumption with IoT - Delivering on Digital Equity and Precision IoT – IoT will Drive the Sweet Spot for Hyper-Connected Consumers in a Hyper-Converged Environment; 2020 dec.

8 Ericsson. Ericsson Mobility Visualizer; 2020. Available from: https://www.ericsson.com/en/mobility-report/mobility-visualizer.

9 5G Americas. 5G - The Future of IoT. 5G Americas White Papers. 2019;(7).

10 Raza U, Kulkarni P, Sooriyabandara M. Low Power Wide Area Networks: An Overview. IEEE Commun Surv Tutorials. 2017;19(2):855–873.

11 AN120.22 LoRa Modulation Basics; 2015.

12 LoRaWAN 1.1 Specification; 2017.

13 Hoeller A, Souza RD, López OLA, Alves H, de Noronha Neto M, Brante G. Analysis and Performance Optimization of LoRa Networks With Time and Antenna Diversity. IEEE Access. 2018;6:32820–32829.

Wireless RF Energy Transfer in the Massive IoT Era: Towards Sustainable Zero-energy Networks.
First Edition. Onel Alcaraz López and Hirley Alves.
© 2022 by The Institute of Electrical and Electronics Engineers, Inc. Published 2022 by John Wiley & Sons, Inc. Companion website: www.wiley.com/go/Alves/WirelessEnergyTransfer

14 Sant'Ana JMdS, Hoeller A, Souza RD, Montejo-Sánchez S, Alves H, Noronha-Neto Md. Hybrid Coded Replication in LoRa Networks. IEEE Trans Ind Informat. 2020;16(8):5577–5585.

15 de Souza Sant'Ana JM, Hoeller A, Souza RD, Alves H, Montejo-Sánchez S. LoRa Performance Analysis with Superposed Signal Decoding. IEEE Wireless Commun Lett. 2020;9(11):1865–1868.

16 de Jesus GGM, Souza RD, Montez C, Hoeller A. LoRaWAN Adaptive Data Rate With Flexible Link Margin. IEEE Internet Things J. 2021;8(7):6053–6061.

17 Hoeller A, Souza RD, Montejo-Sánchez S, Alves H. Performance Analysis of Single-Cell Adaptive Data Rate-Enabled LoRaWAN. IEEE Wireless Commun Lett. 2020;9(6):911–914.

18 Hoeller A, Souza RD, Alves H, López OLA, Montejo-Sánchez S, Pellenz ME. Optimum LoRaWAN Configuration Under Wi-SUN Interference. IEEE Access. 2019;7:170936–170948.

19 Santos Filho FHCd, Dester PS, Stancanelli EMG, Cardieri P, Nardelli PHJ, Carrillo D, et al. Performance of LoRaWAN for Handling Telemetry and Alarm Messages in Industrial Applications. Sensors. 2020;20(11).

20 Shanmuga Sundaram JP, Du W, Zhao Z. A Survey on LoRa Networking: Research Problems, Current Solutions, and Open Issues. IEEE Commun Surveys Tuts. 2020;22(1):371–388.

21 SigFox. SigFox - Technology. Available from: https://www.sigfox.com/en/what-sigfox/technology.

22 Lavric A, Petrariu AI, Popa V. Long Range SigFox Communication Protocol Scalability Analysis Under Large-Scale, High-Density Conditions. IEEE Access. 2019;7:35816–35825.

23 Joris L, Dupont F, Laurent P, Bellier P, Stoukatch S, Redouté JM. An Autonomous Sigfox Wireless Sensor Node for Environmental Monitoring. IEEE Sens Lett. 2019;3(7):01–04.

24 Mikhaylov K, Stusek M, Masek P, Fujdiak R, Mozny R, Andreev S, et al. Communication Performance of a Real-Life Wide-Area Low-Power Network Based on Sigfox Technology. In: ICC ICC; 2020. p. 1–6.

25 Vos G, Al E. Evaluation of LTE-M towards 5G IoT requirements. Sierra Wirel White Pap. (Mar) 2018.

26 Xia N, Chen HH, Yang CS. Emerging Technologies for Machine-Type Communication Networks. IEEE Netw. jan 2020; 34(1):214–222.

27 Mahmood NH, López OLA, Clazzer F, Munari A. Random Access for Cellular Systems. In: Wiley 5G Ref. Wiley; 2020. p. 1–25. Available from: https://onlinelibrary.wiley.com/doi/full/10.1002/9781119471509.w5GRef030.

28 Kanj M, Savaux V, Le Guen M. A Tutorial on NB-IoT Physical Layer Design. IEEE Commun Surv Tutorials. 2020;22(4):2408–2446.

29 Ikpehai A, Adebisi B, Rabie KM, Anoh K, Ande RE, Hammoudeh M, et al. Low-Power Wide Area Network Technologies for Internet-of-Things: A Comparative Review. IEEE Internet Things J. 2019;6(2):2225–2240.

30 GSMA. Extended Coverage –GSM –Internet of Things (EC-GSM-IoT). Available from: https://www.gsma.com/iot/extended-coverage-gsm-internet-of-things-ec-gsm-iot/.

31 Sutton GJ, Zeng J, Liu RP, Ni W, Nguyen DN, Jayawickrama BA, et al. Enabling Technologies for Ultra-Reliable and Low Latency Communications: From PHY and MAC Layer Perspectives. IEEE Commun Surv Tutorials. 2019;21(3):2488–2524.

32 Popovski P, Stefanovic C, Nielsen JJ, de Carvalho E, Angjelichinoski M, Trillingsgaard KF, et al. Wireless Access in Ultra-Reliable Low-Latency Communication (URLLC). IEEE Trans Commun. 2019 aug;67(8):5783–5801.

33 Krishnan A. LoRaWAN and Multi-RAN Architecture Connecting the Next Billion IoT Devices; 2020. Available from: https://info.semtech.com/abi-research-white-paper.

34 International Telecommunication Union. Recommendation ITU-R M.2083-0 IMT Vision – Framework and overall objectives of the future development of IMT for 2020 and beyond. M Series. 2015;9.

35 Mahmood NH, López OLA, Park OS, Moerman I, Mikhaylov K, Mercier E, et al. White paper on critical and massive machine type communication towards 6G. 6G Research Visions. 2020. Available from: http://jultika.oulu.fi/files/isbn9789526226781.pdf.

36 Song H, Bai J, Yi Y, Wu J, Liu L. Artificial Intelligence Enabled Internet of Things: Network Architecture and Spectrum Access. IEEE Comput Intell Mag. 2020 feb;15(1):44–51.

37 Zhu G, Liu D, Du Y, You C, Zhang J, Huang K. Toward an Intelligent Edge: Wireless Communication Meets Machine Learning. IEEE Commun Mag. 2020 jan;58(1):19–25.

38 Ullah MA, Mikhaylov K, Alves H. Massive Machine Type Communication and Satellite Integration for Remote Areas. IEEE Commun Mag. 2021:1–1.

39 Wang P, Zhang J, Zhang X, Yan Z, Evans BG, Wang W. Convergence of Satellite and Terrestrial Networks: A Comprehensive Survey. IEEE Access. 2020;8:5550–5588.

40 Giuliano R, Mazzenga F, Vizzarri A. Satellite-Based Capillary 5G-mMTC Networks for Environmental Applications. IEEE Aerosp Electron Syst Mag. 2019;34(10):40–48.

41 Fraire JA, Céspedes S, Accettura N. Direct-To-Satellite IoT - A Survey of the State of the Art and Future Research Perspectives. In: Palattella MR, Scanzio S, Coleri Ergen S, editors. Ad-Hoc, Mobile, and Wireless Networks. Cham: Springer International Publishing; 2019. p. 241–258.

42 Hexa-X. Deliverable 1.1: 6G Vision, use cases and key societal values; 2021.

43 Matinmikko-Blue M, Aalto S, Asghar MI, Berndt H, Chen Y, Dixit S, et al. White Paper on 6G Drivers and the UN SDGs. 2. University of Oulu; 2020.

44 United Nations Framework Convention on Climate Change. ICT Sector Helping to Tackle Climate Change. Available from: https://unfccc.int/news/ict-sector-helping-to-tackle-climate-change.

45 European Commission. A Europe fit for the digital age. Available from: https://ec.europa.eu/info/strategy/priorities-2019-2024/europe-fit-digital-age_en.

46 European Commission. A European Green Deal. Available from: https://ec.europa.eu/info/strategy/priorities-2019-2024/european-green-deal_en.

47 CCPI. The Climate Change Performance Index Results 2020; 2020.

48 Liikenne- ja viestintäministeriö. Finland announces climate strategy for ICT sector: harnessing data bits to combat climate change. Available from: https://www.lvm.fi/-/finland-announces-climate-strategy-for-ict-sector-harnessing-data-bits-to-combat-climate-change-1260885.

49 Malmodin J, Lundén D. The Energy and Carbon Footprint of the Global ICT and E&M Sectors 2010 – 2015. Sustainability. 2018 aug;10(9):3027.

50 Nokia. People & Planet Report. Nokia Excecutive Rep. 2019:109. https://www.nokia.com/sites/default/files/2020-03/Nokia_People_and_Planet_Report_2019.pdf

51 Nokia. Combating climate change. Available from: https://www.nokia.com/about-us/sustainability/climate/.

52 Ericsson. Climate Action through technology. Available from: https://www.ericsson.com/en/about-us/sustainability-and-corporate-responsibility/environment/climate-action.

53 Wu Q, Zhang R. Towards Smart and Reconfigurable Environment: Intelligent Reflecting Surface Aided Wireless Network. IEEE Commun Mag. 2020;58(1):106–112.

54 Di Renzo M, Debbah M, Phan-Huy DT, Zappone A, Alouini MS, Yuen C, et al. Smart radio environments empowered by reconfigurable AI meta-surfaces: An idea whose time has come. EURASIP J Wirel Commun Netw. 2019;2019(1):1–20.

55 Tasolamprou AC, Pitilakis A, Abadal S, Tsilipakos O, Timoneda X, Taghvaee H, et al. Exploration of Intercell Wireless Millimeter-Wave Communication in the Landscape of Intelligent Metasurfaces. IEEE Access. 2019;7: 122931–122948.

56 Sanchez-Iborra R, Skarmeta AF. TinyML-Enabled Frugal Smart Objects: Challenges and Opportunities. IEEE Circuits Syst Mag. 2020;20(3):4–18.

57 Wang X, Magno M, Cavigelli L, Benini L. FANN-on-MCU: An Open-Source Toolkit for Energy-Efficient Neural Network Inference at the Edge of the Internet of Things. IEEE Internet Things J. 2020 may;7(5):4403–4417.

58 Maraqa O, Rajasekaran AS, Al-Ahmadi S, Yanikomeroglu H, Sait SM. A Survey of Rate-Optimal Power Domain NOMA with Enabling Technologies of Future Wireless Networks. IEEE Commun Surv Tutorials. 2020 oct;22(4):2192–2235.

59 Dai L, Wang B, Ding Z, Wang Z, Chen S, Hanzo L. A Survey of Non-Orthogonal Multiple Access for 5G. IEEE Commun Surv Tutorials. 2018;20(3):2294–2323.

60 Mao Y, Clerckx B, Li VOK. Rate-splitting multiple access for downlink communication systems: bridging, generalizing, and outperforming SDMA and NOMA. EURASIP J Wirel Commun Netw. 2018 dec;2018(1):133.

61 Jaafar W, Naser S, Muhaidat S, Sofotasios PC, Yanikomeroglu H. Multiple Access in Aerial Networks: From Orthogonal and Non-Orthogonal to Rate-Splitting. IEEE Open J Veh Technol. 2020 may;1:372–392.

62 Dizdar O, Mao Y, Han W, Clerckx B. Rate-Splitting Multiple Access: A New Frontier for the PHY Layer of 6G. IEEE 92nd Vehicular Technology Conference (VTC2020-Fall) 2020. 10.1109/VTC2020-Fall49728.2020. 9348672.

63 Wu Y, Gao X, Zhou S, Yang W, Polyanskiy Y, Caire G. Massive Access for Future Wireless Communication Systems. IEEE Wirel Commun. 2020 aug;27(4):148–156.

64 Popovski P, Trillingsgaard KF, Simeone O, Durisi G. 5G Wireless Network Slicing for eMBB, URLLC, and mMTC: A Communication-Theoretic View. IEEE Access. 2018;6:55765–55779.

65 Tang J, Shim B, Quek TQS. Service Multiplexing and Revenue Maximization in Sliced C-RAN Incorporated With URLLC and Multicast eMBB. IEEE J Sel Areas Commun. 2019 apr;37(4):881–895.

66 Kassab R, Simeone O, Popovski P, Islam T. Non-Orthogonal Multiplexing of Ultra-Reliable and Broadband Services in Fog-Radio Architectures. IEEE Access. 2019;7:13035–13049.

67 Tominaga EN, Alves H, López OLA, Souza RD, Rebelatto JL and Latva-aho M. Network Slicing for eMBB and mMTC with NOMA. IEEE 93rd Vehicular Technology Conference (VTC2021-Spring) and Space Diversity Reception, 1-6 2021 10.1109/VTC2021-Spring51267.2021.9448974

68 Tominaga EN, Alves H, López OLA, Souza RD, Rebelatto JL and Latva-aho M. Non-Orthogonal Multiple Access and Network Slicing: Scalable Coexistence of eMBB and URLLC. IEEE 93rd Vehicular Technology Conference (VTC2021-Spring) 1-6 10.1109/VTC2021-Spring51267.2021. 9448942

69 Wang D, Wang J, You X, Wang Y, Chen M, Hou X. Spectral Efficiency of Distributed MIMO Systems. IEEE J Sel Areas Commun. 2013 oct;31(10):2112–2127.

70 Bjornson E, Sanguinetti L. Scalable Cell-Free Massive MIMO Systems. IEEE Trans Commun. 2020 jul;68(7):4247–4261.

71 Wang D, Zhang C, Du Y, Zhao J, Jiang M, You X. Implementation of a Cloud-Based Cell-Free Distributed Massive MIMO System. IEEE Commun Mag. 2020 aug;58(8):61–67.

72 Demir ÖT, Björnson E, Sanguinetti L. Foundations of User-Centric Cell-Free Massive MIMO. Found Trends® Signal Process. 2021;14(3-4):162–472.

73 Ke M, Gao Z, Wu Y, Gao X, Wong KK. Massive Access in Cell-Free Massive MIMO-Based Internet of Things: Cloud Computing and Edge Computing Paradigms. IEEE J Sel Areas Commun. 2020;14(8):1–1.

74 Shannon CE. A mathematical theory of communication. Bell Syst Tech J. 1948;27(3):379–423.

75 Polyanskiy Y, Poor HV, Verdu S. Channel Coding Rate in the Finite Blocklength Regime. IEEE Trans Inf Theory. 2010;56(5):2307–2359.

76 López OLA, Fernández EMG, Souza RD, Alves H. Wireless Powered Communications with Finite Battery and Finite Blocklength. IEEE Trans Commun. 2018;66(4):1803–1816.

77 López OLA, Alves H, Souza RD, Montejo-Sánchez S, García Fernández EM. Rate Control for Wireless-Powered Communication Network With Reliability and Delay Constraints. IEEE Trans Wireless Commun. 2019;18(12):5791–5805.

78 López OLA, Alves H, Souza RD, Montejo-Sánchez S, Fernandez EMG, Latva-aho M. Massive Wireless Energy Transfer: Enabling Sustainable IoT Towards 6G Era. IEEE Internet Things J. 2021;8(11)8816–8835. 10.1109/JIOT.2021.3050612

79 Haight R, Haensch W, Friedman D. Solar-powering the Internet of Things. Science (80-). 2016;353(6295):124–126.

80 Haque T, Elkotby H, Cabrol P, Pragada R, Castor D. A Supplemental Zero-Energy Downlink Air-Interface Enabling 40-Year Battery Life in IoT Devices. In: GLOBECOM. vol. 2020-Janua. IEEE; 2020. p. 1–6.

81 United Nations Environment Management Group. United Nations System-wide Response to Tackling E-waste. 2017:60. Available from: https://unemg.org/images/emgdocs/ewaste/E-Waste-EMG-FINAL.pdf.

82 Global E-Waste Surging: Up 21% in 5 Years - United Nations University. Available from: https://unu.edu/media-relations/releases/global-e-waste-surging-up-21-in-5-years.html.

83 Lithium-Ion Battery Recycling Finally Takes Off in North America and Europe - IEEE Spectrum. Available from: https://spectrum.ieee.org/energy/batteries-storage/lithiumion-battery-recycling-finally-takes-off-in-north-america-and-europe.

84 López OLA, Alves H, Souza RD, Montejo-Sánchez S. Statistical Analysis of Multiple Antenna Strategies for Wireless Energy Transfer. IEEE Trans Commun. 2019;67(10):7245–7262.

85 López OLA, Monteiro FA, Alves H, Zhang R, Latva-aho M. A Low-Complexity Beamforming Design for Multiuser Wireless Energy

Transfer. IEEE Wireless Commun Lett. 2021;10(1):58–62.
10.1109/LWC.2020.3020576

86 López OLA, Mahmood NH, Alves H, Lima CM, Latva-aho M. Ultra-Low
Latency, Low Energy, and Massiveness in the 6G Era via Efficient
CSIT-Limited Scheme. IEEE Commun Mag. 2020;58(11):56–61.

87 López OLA, Montejo-Sánchez S, Souza RD, Papadias CB, Alves H. On
CSI-free Multi-Antenna Schemes for Massive RF Wireless Energy Transfer.
IEEE Internet Things J. 2021;8(1):278–296. 10.1109/JIOT.2020.3003114

88 Rosabal OM, López OLA, Alves H, Montejo-Sánchez S, Latva-aho M. On
the Optimal Deployment of Power Beacons for Massive Wireless Energy
Transfer. IEEE Internet Things J. 2021;8(13):10531–10542.
10.1109/JIOT.2020.3048065

89 Huang K, Zhou X. Cutting the last wires for mobile communications by
microwave power transfer. IEEE Commun Mag. 2015;53(6):86–93.

90 Awal MR, Jusoh M, Sabapathy T, Kamarudin MR, Rahim RA.
State-of-the-art developments of acoustic energy transfer. Int J Antenn
Propag. 2016.

91 Chen Y. Energy Sources. In: Energy Harvesting Communications: Principles
and Theories. John Wiley & Sons; 2019.

92 Prauzek M, Konecny J, Borova M, Janosova K, Hlavica J, Musilek P. Energy
Harvesting Sources, Storage Devices and System Topologies for
Environmental Wireless Sensor Networks: A Review. Sensors. 2018;18(8).

93 Jin K, Zhou W. Wireless Laser Power Transmission: A Review of Recent
Progress. IEEE Trans Power Electron. 2019;34(4):3842–3859.

94 Zhang Q, Fang W, Liu Q, Wu J, Xia P, Yang L. Distributed Laser Charging:
A Wireless Power Transfer Approach. IEEE Internet Things J.
2018;5(5):3853–3864.

95 Valenta CR, Durgin GD. Harvesting Wireless Power: Survey of
Energy-Harvester Conversion Efficiency in Far-Field, Wireless Power
Transfer Systems. IEEE Microw Mag. 2014;15(4):108–120.

96 Divakaran SK, Krishna DD. RF energy harvesting systems: An overview and
design issues. Int J RF Microw C E. 2019;29(1):e21633.

97 Xu X, Özçelikkale A, McKelvey T, Viberg M. Simultaneous information and
power transfer under a non-linear RF energy harvesting model. In: IEEE ICC
Workshops; 2017. p. 179–184.

98 Chen Y, Zhao N, Alouini M. Wireless Energy Harvesting Using Signals From
Multiple Fading Channels. IEEE Trans Commun. 2017;65(11):5027–5039.

99 Chen Y, Sabnis KT, Abd-Alhameed RA. New Formula for Conversion
Efficiency of RF EH and Its Wireless Applications. IEEE Trans Veh Technol.
2016;65(11):9410–9414.

100 Boshkovska E, Ng DWK, Zlatanov N, Schober R. Practical Non-Linear
Energy Harvesting Model and Resource Allocation for SWIPT Systems.
IEEE Commun Lett. 2015;19(12):2082–2085.

101 Alevizos PN, Bletsas A. Sensitive and Nonlinear Far-Field RF Energy Harvesting in Wireless Communications. IEEE Trans Wireless Commun. 2018;17(6):3670–3685.

102 Zhu P, Ma Z, Vandenbosch GAE, Gielen G. 160 GHz harmonic-rejecting antenna with CMOS rectifier for millimeter-wave wireless power transmission. In: 9th EuCAP; 2015. p. 1–5.

103 Zhang J, Wu ZP, Liu CG, Zhang BH, Zhang B. A double-sided rectenna design for RF energy harvesting. In: IEEE IWS; 2015. p. 1–4.

104 Lu P, Yang X, Li J, Wang B. A Compact Frequency Reconfigurable Rectenna for 5.2- and 5.8-GHz Wireless Power Transmission. IEEE Trans Power Electron. 2015;30(11):6006–6010.

105 Matsunaga T, Nishiyama E, Toyoda I. 5.8-GHz Stacked Differential Rectenna Suitable for Large-Scale Rectenna Arrays With DC Connection. IEEE Trans Antennas Propag. 2015;63(12):5944–5949.

106 Song C, Huang Y, Zhou J, Zhang J, Yuan S, Carter P. A High-Efficiency Broadband Rectenna for Ambient Wireless Energy Harvesting. IEEE Trans Antennas Propag. 2015;63(8):3486–3495.

107 Momenroodaki P, Fernandes RD, Popović Z. Air-substrate compact high gain rectennas for low RF power harvesting. In: 10th EuCAP; 2016. p. 1–4.

108 Lu P, Yang X, Li J, Wang B. Polarization Reconfigurable Broadband Rectenna with Tunable Matching Network for Microwave Power Transmission. IEEE Trans Antennas Propag. 2016;64(3):1136–1141.

109 Sun H. An Enhanced Rectenna using Differentially-Fed Rectifier for Wireless Power Transmission. IEEE Antennas Wireless Propag Lett. 2016;15:32–35.

110 Sun H, Geyi W. A New Rectenna with All-Polarization-Receiving Capability for Wireless Power Transmission. IEEE Antennas Wireless Propag Lett. 2016;15:814–817.

111 AbdelTawab AM, Khattab A. Efficient multi-band energy harvesting circuit for Wireless Sensor nodes. In: Fourth International Japan-Egypt Conference on Electronics, Communications and Computers (JEC-ECC); 2016. p. 75–78.

112 Dastranj A. Very small planar broadband monopole antenna with hybrid trapezoidal–elliptical radiator. IET Microw Antennas Propag. 2016;11(4):542–547.

113 Song C, Huang Y, Zhou J, Carter P, Yuan S, Xu Q, et al. Matching Network Elimination in Broadband Rectennas for High-Efficiency Wireless Power Transfer and Energy Harvesting. IEEE Trans Ind Electron. 2017;64(5):3950–3961.

114 Agrawal S, Gupta RD, Parihar MS, Kondekar PN. A wideband high gain dielectric resonator antenna for RF energy harvesting application. AEU-International Journal of Electronics and Communications. 2017;78:24–31.

115 Suri K, Mohta M, Rajawat A, Gupta SH. A 5.8 GHz inset-fed rectenna for RF energy harvesting applications. In: ICCT; 2017. p. 32–35.

116 Salsabila S, Munir A. 1.8GHz printed bow-tie dipole rectenna with voltage quadrupler for RF energy harvesting. In: TENCON; 2017. p. 2739–2742.

117 Palazzi V, Hester J, Bito J, Alimenti F, Kalialakis C, Collado A, et al. A Novel Ultra-Lightweight Multiband Rectenna on Paper for RF Energy Harvesting in the Next Generation LTE Bands. IEEE Trans Microw Theory Techn. 2018;66(1):366–379.

118 Shen S, Chiu C, Murch RD. Multiport Pixel Rectenna for Ambient RF Energy Harvesting. IEEE Trans Antennas Propag. 2018;66(2):644–656.

119 Muncuk U, Alemdar K, Sarode JD, Chowdhury KR. Multiband Ambient RF Energy Harvesting Circuit Design for Enabling Batteryless Sensors and IoT. IEEE Internet Things J. 2018;5(4):2700–2714.

120 Chandravanshi S, Sarma SS, Akhtar MJ. Design of Triple Band Differential Rectenna for RF Energy Harvesting. IEEE Trans Antennas Propag. 2018;66(6):2716–2726.

121 Amjad O, Munir SW, Imeci ST, Ercan AÖ. Design and implementation of dual band microstrip patch antenna for WLAN energy harvesting system. Appl Comput Electromagn Soc J. 2018.

122 Partal HP, Belen MA, Partal SZ. Design and realization of an ultra-low power sensing RF energy harvesting module with its RF and DC sub-components. Int J RF Microw C E. 2019;29(1):e21622.

123 Palandoken M, Gocen C. A modified Hilbert fractal resonator based rectenna design for GSM900 band RF energy harvesting applications. Int J RF Microw C E. 2019;29(1):e21643.

124 Zhu L, Zhang J, Han W, Xu L, Bai X. A novel RF energy harvesting cube based on air dielectric antenna arrays. Int J RF Microw C E. 2019;29(1):e21636.

125 Sampe J, Yunus NHM, Yunas J, Pawi A. Architecture of an efficient dual band 1.8/2.5 GHz rectenna for RF energy harvesting. Telkomnika. 2019;17(6):3137–3144.

126 Colaiuda D, Ulisse I, Ferri G. Rectifiers' Design and Optimization for a Dual-Channel RF Energy Harvester. J Low Power Electron Appl. 2020;10(2):11.

127 Mohd Yunus NH, Sampe J, Yunas J, Pawi A, Rhazali ZA. MEMS based antenna of energy harvester for wireless sensor node. Microsyst Technol:1–8.

128 Elwi TA, Jassim DA, Mohammed HH. Novel miniaturized folded UWB microstrip antenna-based metamaterial for RF energy harvesting. Int J Commun Syst. 2020;33(6):e4305.

129 Singh N, Kumar S, Kanaujia BK, Beg MT, Kumar S, et al. A compact and efficient graphene FET based RF energy harvester for green communication. AEU. 2020;115:153059.

130 Elsheakh D, Farouk M, Elsadek H, Ghali H. Quad-Band Rectenna for RF Energy Harvesting System. JEMAA. 2020;12(5):57–70.

131 Lu X, Wang P, Niyato D, Kim DI, Han Z. Wireless Networks With RF Energy Harvesting: A Contemporary Survey. IEEE Commun Surveys Tuts. 2015;17(2):757–789.

132 Tran H, Kaddoum G. RF Wireless Power Transfer: Regreening Future Networks. IEEE Potentials. 2018;37(2):35–41.

133 Liu X, Ansari N. Toward Green IoT: Energy Solutions and Key Challenges. IEEE Commun Mag. 2019;57(3):104–110.

134 Jiang L, Xie S, Maharjan S, Zhang Y. Blockchain Empowered Wireless Power Transfer for Green and Secure Internet of Things. IEEE Netw. 2019;33(6):164–171.

135 Soyata T, Copeland L, Heinzelman W. RF energy harvesting for embedded systems: A survey of tradeoffs and methodology. IEEE Circuits Syst Mag. 2016;16(1):22–57.

136 Zhang Y, Zhang F, Shakhsheer Y, Silver JD, Klinefelter A, Nagaraju M, et al. A Batteryless 19 μW MICS/ISM-Band Energy Harvesting Body Sensor Node SoC for ExG Applications. IEEE J Solid-State Circuits. 2013;48(1):199–213.

137 Cao S, Li J. A survey on ambient energy sources and harvesting methods for structural health monitoring applications. Adv Mech Eng. 2017;9(4):1687814017696210.

138 Saraereh OA, Alsaraira A, Khan I, Choi BJ. A hybrid energy harvesting design for on-body Internet-of-Things (IoT) networks. Sensors. 2020;20(2):407.

139 Tran LG, Cha HK, Park WT. RF power harvesting: a review on designing methodologies and applications. Micro and Nano Systems Letters. 2017;5(1):14.

140 Carrillo D, Mikhaylov K, Nardelli PJ, Andreev S, da Costa DB. Understanding UAV-Based WPCN-Aided Capabilities for Offshore Monitoring Applications. IEEE Wireless Commun. 2021:1–7.

141 Ruisi Ge, Hong Pan, Zhibin Lin, Na Gong, Jinhui Wang, Xiaowei Chen. RF-powered battery-less Wireless Sensor Network in structural monitoring. In: IEEE EIT; 2016. p. 0547–0552.

142 Clerckx B, Zhang R, Schober R, Ng DWK, Kim DI, Poor HV. Fundamentals of Wireless Information and Power Transfer: From RF Energy Harvester Models to Signal and System Designs. IEEE J Sel Areas Commun. 2019;37(1):4–33.

143 Alsaba Y, Rahim SKA, Leow CY. Beamforming in Wireless Energy Harvesting Communications Systems: A Survey. IEEE Commun Surveys Tuts. 2018;20(2):1329–1360.

144 Choi KW, Aziz AA, Setiawan D, Tran NM, Ginting L, Kim DI. Distributed Wireless Power Transfer System for Internet of Things Devices. IEEE Internet Things J. 2018;5(4):2657–2671.

145 Interdonato G, Björnson E, Ngo HQ, Frenger P, Larsson EG. Ubiquitous cell-free massive MIMO communications. EURASIP J Wirel Commun Netw. 2019;2019(1):197.

146 Cansiz M, Altinel D, Kurt GK. Efficiency in RF energy harvesting systems: A comprehensive review. Energy. 2019;174:292–309.

147 Zhang Q, Li F, Wang Y. Mobile Crowd Wireless Charging Toward Rechargeable Sensors for Internet of Things. IEEE Internet Things J. 2018;5(6):5337–5347.

148 Wang N, Wu J, Dai H. Bundle Charging: Wireless Charging Energy Minimization in Dense Wireless Sensor Networks. In: IEEE 39th ICDCS; 2019. p. 810–820.

149 Dai H, Wang X, Xu L, Dong C, Liu Q, Meng L, et al. Area Charging for Wireless Rechargeable Sensors. In: 29th ICCCN; 2020. p. 1–9.

150 Liu Y, Dai HN, Wang Q, Imran M, Guizani N. Wireless Powering Internet of Things with UAVs: Challenges and Opportunities. arXiv preprint arXiv:200905220. 2020.

151 Hussaini A, Al-Yasir Y, Voudouris K, Abd-Alhameed R, Husham M, Elfergani I, et al. In: Green Flexible RF for 5G; 2015. p. 241–272.

152 Wolfe S, Begashaw S, Liu Y, Dandekar KR. Adaptive Link Optimization for 802.11 UAV Uplink Using a Reconfigurable Antenna. In: MILCOM; 2018. p. 1–6.

153 Zeng T, Mozaffari M, Semiari O, Saad W, Bennis M, Debbah M. Wireless Communications and Control for Swarms of Cellular-Connected UAVs. In: 52nd ACSSC; 2018. p. 719–723.

154 Lyu B, Hoang DT, Gong S, Yang Z. Intelligent Reflecting Surface Assisted Wireless Powered Communication Networks. In: IEEE WCNCW; 2020. p. 1–6.

155 Tang Y, Ma G, Xie H, Xu J, Han X. Joint Transmit and Reflective Beamforming Design for IRS-Assisted Multiuser MISO SWIPT Systems. In: IEEE ICC; 2020. p. 1–6.

156 Pan C, Ren H, Wang K, Elkashlan M, Nallanathan A, Wang J, et al. Intelligent Reflecting Surface Aided MIMO Broadcasting for Simultaneous Wireless Information and Power Transfer. IEEE J Sel Areas Commun. 2020;38(8):1719–1734.

157 López OLA, Alves H. In: Alves H, Riihonen T, Suraweera HA, editors. Full Duplex and Wireless-Powered Communications. Singapore: Springer Singapore; 2020. p. 219–248. Available from: https://doi.org/10.1007/978-981-15-2969-6_8.

158 Portilla J, Mujica G, Lee J, Riesgo T. The Extreme Edge at the Bottom of the Internet of Things: A Review. IEEE Sensors J. 2019;19(9):3179–3190.

159 Tabesh M, Dolatsha N, Arbabian A, Niknejad AM. A Power-Harvesting Pad-Less Millimeter-Sized Radio. IEEE J Solid-State Circuits. 2015;50(4):962–977.

160 Hampton JR. Introduction to MIMO communications. Cambridge Univ. Press; 2014.

161 Bi S, Ho CK, Zhang R. Wireless powered communication: Opportunities and challenges. IEEE Commun Mag. 2015 April;53(4):117–125.

162 Ju H, Zhang R. Throughput Maximization in Wireless Powered Communication Networks. IEEE Trans Wireless Commun. 2014;13(1):418–428.

163 Zhang R, Ho CK. MIMO Broadcasting for Simultaneous Wireless Information and Power Transfer. IEEE Trans Wireless Commun. 2013;12(5):1989–2001.

164 López OLA, Fernández EMG, Souza RD, Alves H. Ultra-Reliable Cooperative Short-Packet Communications With Wireless Energy Transfer. IEEE Sensors J. 2018;18(5):2161–2177.

165 López OLA, Demo Souza R, Alves H, Martín García Fernández E. Ultra reliable short message relaying with wireless power transfer. In: IEEE ICC; 2017. p. 1–6.

166 Zohdy M, ElBatt T, Nafie M, Ercetin O. RF Energy Harvesting in Wireless Networks with HARQ. In: IEEE GC Wkshps; 2016. p. 1–6.

167 Abad MSH, Ercetin O, Elbatt T, Nahe M. Wireless energy and information transfer in networks with hybrid ARQ. In: IEEE WCNC; 2018. p. 1–6.

168 Li Z, Kang J, Yu R, Ye D, Deng Q, Zhang Y. Consortium Blockchain for Secure Energy Trading in Industrial Internet of Things. IEEE Trans Ind Informat. 2018;14(8):3690–3700.

169 Sample A, Smith JR. Experimental results with two wireless power transfer systems. In: IEEE RWS; 2009. p. 16–18.

170 Shigeta R, Sasaki T, Quan DM, Kawahara Y, Vyas RJ, Tentzeris MM, et al. Ambient RF Energy Harvesting Sensor Device with Capacitor-Leakage-Aware Duty Cycle Control. IEEE Sensors J. 2013;13(8):2973–2983.

171 Lee S, Zhang R, Huang K. Opportunistic Wireless Energy Harvesting in Cognitive Radio Networks. IEEE Trans Wireless Commun. 2013;12(9):4788–4799.

172 Georgiou O, Mimis K, Halls D, Thompson WH, Gibbins D. How Many Wi-Fi APs Does it Take to Light a Lightbulb? IEEE Access. 2016;4:3732–3746.

173 Stoopman M, Keyrouz S, Visser HJ, Philips K, Serdijn WA. Co-Design of a CMOS Rectifier and Small Loop Antenna for Highly Sensitive RF Energy Harvesters. IEEE J Solid-State Circuits. 2014;49(3):622–634.

174 Papotto G, Carrara F, Finocchiaro A, Palmisano G. A 90-nm CMOS 5-Mbps Crystal-Less RF-Powered Transceiver for Wireless Sensor Network Nodes. IEEE J Solid-State Circuits. 2014;49(2):335–346.

175 Rosa RL, Zoppi G, Finocchiaro A, Papotto G, Di Donato L, Sorbello G, et al. An over-the-distance wireless battery charger based on RF energy harvesting. In: 14th SMACD; 2017. p. 1–4.

176 Wang X, Mortazawi A. Medium Wave Energy Scavenging for Wireless Structural Health Monitoring Sensors. IEEE Trans Microw Theory Techn. 2014;62(4):1067–1073.

177 Leon-Gil JA, Cortes-Loredo A, Fabian-Mijangos A, Martinez-Flores JJ, Tovar-Padilla M, Cardona-Castro MA, et al. Medium and Short Wave RF Energy Harvester for Powering Wireless Sensor Networks. Sensors. 2018;18(3).

178 Kwan JC, Fapojuwo AO. Radio Frequency Energy Harvesting and Data Rate Optimization in Wireless Information and Power Transfer Sensor Networks. IEEE Sensors J. 2017;17(15):4862–4874.

179 Ku M, Li W, Chen Y, Ray Liu KJ. Advances in Energy Harvesting Communications: Past, Present, and Future Challenges. IEEE Commun Surveys Tuts. 2016;18(2):1384–1412.

180 Liu J, Dai H, Chen W. Delay Optimal Scheduling for Energy Harvesting Based Communications. IEEE J Sel Areas Commun. 2015;33(3):452–466.

181 Xu J, Duan L, Zhang R. Cost-aware green cellular networks with energy and communication cooperation. IEEE Commun Mag. 2015;53(5):257–263.

182 Ghazanfari A, Tabassum H, Hossain E. Ambient RF energy harvesting in ultra-dense small cell networks: performance and trade-offs. IEEE Wireless Commun. 2016;23(2):38–45.

183 Niotaki K, Kim S, Jeong S, Collado A, Georgiadis A, Tentzeris MM. A Compact Dual-Band Rectenna Using Slot-Loaded Dual Band Folded Dipole Antenna. IEEE Antennas Wireless Propag Lett. 2013;12:1634–1637.

184 Zeng M, Andrenko AS, Liu X, Li Z, Tan H. A Compact Fractal Loop Rectenna for RF Energy Harvesting. IEEE Antennas Wireless Propag Lett. 2017;16:2424–2427.

185 Kimionis J, Isakov M, Koh BS, Georgiadis A, Tentzeris MM. 3D-Printed Origami Packaging With Inkjet-Printed Antennas for RF Harvesting Sensors. IEEE Trans Microw Theory Techn. 2015;63(12):4521–4532.

186 Vandelle E, Bui DHN, Vuong T, Ardila G, Wu K, Hemour S. Harvesting Ambient RF Energy Efficiently With Optimal Angular Coverage. IEEE Trans Antennas Propag. 2019;67(3):1862–1873.

187 Vandelle E, Doan PL, Bui DHN, Vuong TP, Ardila G, Wu K, et al. High gain isotropic rectenna. In: IEEE WPTC; 2017. p. 1–4.

188 Lee D, Lee S, Hwang I, Lee W, Yu J. Hybrid Power Combining Rectenna Array for Wide Incident Angle Coverage in RF Energy Transfer. IEEE Trans Microw Theory Techn. 2017;65(9):3409–3418.

189 Wu Q, Zhang S, Zheng B, You C, Zhang R. Intelligent Reflecting Surface-Aided Wireless Communications: A Tutorial. IEEE Trans Commun. 2021;69(5):3313–3351.

190 Yuan X, Zhang YJA, Shi Y, Yan W, Liu H. Reconfigurable-Intelligent-Surface Empowered Wireless Communications: Challenges and Opportunities. IEEE Wireless Commun. 2021;28(2):136–143.

191 Alexandropoulos GC, Lerosey G, Debbah M, Fink M. Reconfigurable Intelligent Surfaces and Metamaterials: The Potential of Wave Propagation Control for 6G Wireless Communications. arXiv preprint arXiv:200611136. 2020.

192 Vullers R, Visser H, het Veld BO, Pop V. RF harvesting using antenna structures on foil. Proc PowerMEMS. 2008:9–12.

193 Erkmen F, Almoneef TS, Ramahi OM. Scalable Electromagnetic Energy Harvesting Using Frequency-Selective Surfaces. IEEE Trans Microw Theory Techn. 2018;66(5):2433–2441.

194 Vital D, Bhardwaj S, Volakis JL. Textile-Based Large Area RF-Power Harvesting System for Wearable Applications. IEEE Trans Antennas Propag. 2020;68(3):2323–2331.

195 Collado A, Georgiadis A. Conformal Hybrid Solar and Electromagnetic (EM) Energy Harvesting Rectenna. IEEE Trans Circuits Syst I, Reg Papers. 2013;60(8):2225–2234.

196 Roselli L, Borges Carvalho N, Alimenti F, Mezzanotte P, Orecchini G, Virili M, et al. Smart Surfaces: Large Area Electronics Systems for Internet of Things Enabled by Energy Harvesting. Proc IEEE. 2014;102(11):1723–1746.

197 Sun K, Fan R, Zhang X, Zhang Z, Shi Z, Wang N, et al. An overview of metamaterials and their achievements in wireless power transfer. J Mater Chem C. 2018;6:2925–2943.

198 Piñuela M, Mitcheson PD, Luçyszyn S. Ambient RF Energy Harvesting in Urban and Semi-Urban Environments. IEEE Trans Microw Theory Techn. 2013;61(7):2715–2726.

199 Cansiz M, Abbasov T, Kurt MB, Celik AR. Mobile measurement of radiofrequency electromagnetic field exposure level and statistical analysis. Measurement. 2016;86:159–164.

200 Mazloum T, Aerts S, Joseph W, Wiart J. RF-EMF exposure induced by mobile phones operating in LTE small cells in two different urban cities. Ann Telecommun. 2019;74(1-2):35–42.

201 Velghe M, Joseph W, Debouvere S, Aminzadeh R, Martens L, Thielens A. Characterisation of spatial and temporal variability of RF-EMF exposure levels in urban environments in Flanders, Belgium. Environ Res. 2019;175:351–366.

202 Tentzeris MM, Kawahara Y. Novel Energy Harvesting Technologies for ICT Applications. In: International Symposium on Applications and the Internet; 2008. p. 373–376.

203 Bouchouicha D, Dupont F, Latrach M, Ventura L. Ambient RF energy harvesting. In: ICREPQ. vol. 13; 2010. p. 2–6.

204 Mekid S, Qureshi A, Baroudi U. Energy harvesting from ambient radio frequency: Is it worth it? AJSE. 2017;42(7):2673–2683.

205 Huang K. Spatial Throughput of Mobile Ad Hoc Networks Powered by Energy Harvesting. IEEE Trans Inf Theory. 2013;59(11):7597–7612.

206 Chung W, Park S, Lim S, Hong D. Spectrum Sensing Optimization for Energy-Harvesting Cognitive Radio Systems. IEEE Trans Wireless Commun. 2014;13(5):2601–2613.

207 Haenggi M. Stochastic geometry for wireless networks. Cambridge University Press; 2012.

208 Sakr AH, Hossain E. Analysis of *K*-Tier Uplink Cellular Networks With Ambient RF Energy Harvesting. IEEE J Sel Areas Commun. 2015;33(10):2226–2238.

209 Wang L, Liao X, Li Y. Transmission Strategy for D2D Terminal with Ambient RF Energy Harvesting. In: IEEE 86th VTC-Fall; 2017. p. 1–5.

210 Deng N, Haenggi M. The Energy and Rate Meta Distributions in Wirelessly Powered D2D Networks. IEEE J Sel Areas Commun. 2019;37(2):269–282.

211 Kusaladharma S, Tellambura C and Zhang Z. Evaluation of RF Energy Harvesting by Mobile D2D Nodes Within a Stochastic Field of Base Stations. IEEE Transactions on Green Communications and Networking, 2020;4(4): 1120–1129. 10.1109/TGCN.2020.2994513

212 Deng N, Zhou W, Haenggi M. The Ginibre Point Process as a Model for Wireless Networks With Repulsion. IEEE Trans Wireless Commun. 2015;14(1):107–121.

213 Flint I, Lu X, Privault N, Niyato D, Wang P. Performance Analysis of Ambient RF Energy Harvesting with Repulsive Point Process Modeling. IEEE Trans Wireless Commun. 2015;14(10):5402–5416.

214 Lu X, Flint I, Niyato D, Privault N, Wang P. Self-Sustainable Communications With RF Energy Harvesting: Ginibre Point Process Modeling and Analysis. IEEE J Sel Areas Commun. 2016;34(5):1518–1535.

215 Ammar HA, Nasser Y, Artail H. Closed Form Expressions for the Probability Density Function of the Interference Power in PPP Networks. In: IEEE ICC; 2018. p. 1–6.

216 Anderssen RS, Husain SA, Loy RJ. The Kohlrausch function: properties and applications. In: Crawford J, Roberts AJ, editors. Proc. of 11th Computational Techniques and Applications Conference CTAC-2003. vol. 45; 2004. p. C800–C816.

217 Haenggi M. The Meta Distribution of the SIR in Poisson Bipolar and Cellular Networks. IEEE Trans Wireless Commun. 2016;15(4):2577–2589.

218 Deng N, Haenggi M. A Fine-Grained Analysis of Millimeter-Wave Device-to-Device Networks. IEEE Trans Commun. 2017;65(11):4940–4954.

219 Thudugalage A, Atapattu S, Evans J. Beamformer design for wireless energy transfer with fairness. In: IEEE ICC; 2016. p. 1–6.

220 Boyd S, Boyd SP, Vandenberghe L. Convex optimization. Cambridge university press; 2004.

221 Ye Y. Interior point algorithms: theory and analysis. vol. 44. John Wiley & Sons; 2011.

222 Zeng Y, Zhang R. Optimized Training Design for Wireless Energy Transfer. IEEE Trans Commun. 2015 Feb;63(2):536–550.

223 Zeng Y, Zhang R. Optimized Training for Net Energy Maximization in Multi-Antenna Wireless Energy Transfer Over Frequency-Selective Channel. IEEE Trans Commun. 2015 Jun;63(6):2360–2373.

224 Proakis JG. Digital communications. 4th ed. McGraw-Hill; 2001.

225 Khalighi MA, Brossier J, Jourdain G, Raoof K. On capacity of Rician MIMO channels. In: 12th IEEE PIMRC. vol. 1; 2001. p. A–A.

226 Goldsmith A. Wireless communications. Cambridge Univ. Press; 2005.

227 López OLA, Alves H, Montejo-Sánchez S, Souza RD, Latva-aho M. CSI-free Rotary Antenna Beamforming for Massive RF Wireless Energy Transfer. IEEE Internet of Things Journal. 2021: 1–1 10.1109/JIOT.2021.3107222.

228 Schreier PJ, Scharf LL. Statistical signal processing of complex-valued data: the theory of improper and noncircular signals. Cambridge university press; 2010.

229 Thompson I. NIST Handbook of Mathematical Functions. Taylor & Francis; 2011.

230 Clerckx B, Kim J. On the Beneficial Roles of Fading and Transmit Diversity in Wireless Power Transfer with Nonlinear Energy Harvesting. IEEE Trans Wireless Commun. 2018;17(11):7731–7743.

231 Varasteh M, Rassouli B, Clerckx B. On capacity-achieving distributions for complex AWGN channels under nonlinear power constraints and their applications to SWIPT. IEEE Trans Inf Theory. 2020;66(10): 6488–6508 10.1109/TIT.2020.2998464

232 Khan D, Oh SJ, Shehzad K, Verma D, Khan ZHN, Pu YG, et al. A CMOS RF Energy Harvester With 47 Peak Efficiency Using Internal Threshold Voltage Compensation. IEEE Microw Wireless Compon Lett. 2019;29(6):415–417.

233 Botkin ND, Turova-Botkina V. An algorithm for finding the Chebyshev center of a convex polyhedron. Appl Math Optim. 1994;29(2):211–222.

234 Grant M, Boyd S. CVX: Matlab Software for Disciplined Convex Programming, version 2.1; 2014. http://cvxr.com/cvx.

235 Stephenson K. Circle packing: a mathematical tale. Not Am Math Soc. 2003;50(11):1376–1388.

236 Mary P, Gorce J, Unsal A, Poor HV. Finite Blocklength Information Theory: What is the Practical Impact on Wireless Communications? In: IEEE GC Wkshps; 2016. p. 1–6.

237 López OLA, Alves H, Souza RD, Latva-Aho M. Finite Blocklength Error Probability Distribution for Designing Ultra Reliable Low Latency Systems. IEEE Access. 2020;8:107353–107363.

238 Olver FW, Lozier DW, Boisvert RF, Clark CW. NIST handbook of mathematical functions. vol. 521140633. Cambridge university press; 2010.

239 Isikman AO, Yuksel M, Gündüz D. A Low-Complexity Policy for Outage Probability Minimization With an Energy Harvesting Transmitter. IEEE Commun Lett. 2017;21(4):917–920.

240 López OLA, Mahmood NH, Alves H, Latva-aho M. CSI-free vs CSI-based multi-antenna WET for massive low-power Internet of Things. IEEE Trans Wireless Commun. 2021;20(5): 3078–3094. 10.1109/TWC.2020.3047355

241 Tzeng SS, Lin YJ. Wireless energy transfer policies for cognitive radio based MAC in energy-constrained IoT networks. Telecommun Syst. 2021:1–15.

242 Aijaz A, Aghvami AH. Cognitive Machine-to-Machine Communications for Internet-of-Things: A Protocol Stack Perspective. IEEE Internet Things J. 2015;2(2):103–112.

243 Krikidis I, Timotheou S, Nikolaou S, Zheng G, Ng DWK, Schober R. Simultaneous wireless information and power transfer in modern communication systems. IEEE Commun Mag. 2014;52(11):104–110.

244 Xu J, Liu L, Zhang R. Multiuser MISO Beamforming for Simultaneous Wireless Information and Power Transfer. IEEE Trans Signal Process. 2014;62(18):4798–4810.

245 Hunger R, Joham M. A Complete Description of the QoS Feasibility Region in the Vector Broadcast Channel. IEEE Trans Signal Process. 2010;58(7):3870–3878.

246 Luo Z, Ma W, So AM, Ye Y, Zhang S. Semidefinite Relaxation of Quadratic Optimization Problems. IEEE Signal Process Mag. 2010;27(3):20–34.

247 Bengtsson M, Ottersten B. Optimal and suboptimal transmit beamforming. 2001.

248 Wiesel A, Eldar YC, Shamai S. Linear precoding via conic optimization for fixed MIMO receivers. IEEE Trans Signal Process. 2006;54(1):161–176.

249 Sanguinetti L, Björnson E, Debbah M, Moustakas AL. Optimal linear precoding in multi-user MIMO systems: A large system analysis. In: IEEE GLOBECOM. IEEE; 2014. p. 3922–3927.

250 Shi Q, Liu L, Xu W, Zhang R. Joint Transmit Beamforming and Receive Power Splitting for MISO SWIPT Systems. IEEE Trans Wireless Commun. 2014;13(6):3269–3280.

251 Trotter MS, Griffin JD, Durgin GD. Power-optimized waveforms for improving the range and reliability of RFID systems. In: IEEE Int. Conf. on RFID; 2009. p. 80–87.

252 Boaventura AS, Carvalho NB. Maximizing DC power in energy harvesting circuits using multisine excitation. In: IEEE MTT-S Int. Microwave Symp.; 2011. p. 1–4.

253 Clerckx B, Bayguzina E, Yates D, Mitcheson PD. Waveform optimization for Wireless Power Transfer with nonlinear energy harvester modeling. In: ISWCS; 2015. p. 276–280.

254 Clerckx B, Bayguzina E. Waveform Design for Wireless Power Transfer. IEEE Trans Signal Process. 2016;64(23):6313–6328.

255 Clerckx B, Bayguzina E. Low-Complexity Adaptive Multisine Waveform Design for Wireless Power Transfer. IEEE Antennas Wireless Propag Lett. 2017;16:2207–2210.

256 Clerckx B. Wireless information and power transfer: Nonlinearity, waveform design, and rate-energy tradeoff. IEEE Trans Signal Process. 2017;66(4):847–862.

257 Clerckx B, Huang K, Varshney LR, Ulukus S, Alouini MS. Wireless Power Transfer for Future Networks: Signal Processing, Machine Learning, Computing, and Sensing; IEEE Journal of Selected Topics in Signal Processing. 2021: 1–1 10.1109/JSTSP.2021.3098478

258 Smith JG. The information capacity of amplitude-and variance-constrained scalar Gaussian channels. Inf Control. 1971;18(3):203–219.

259 Shamai S, Bar-David I. The capacity of average and peak-power-limited quadrature Gaussian channels. IEEE Trans Inf Theory. 1995;41(4): 1060–1071.

260 Zhou X, Zhang R, Ho CK. Wireless information and power transfer in multiuser OFDM systems. IEEE Trans Wireless Commun. 2014;13(4):2282–2294.

261 Demir OT, Tuncer TE. Max – Min Fair Resource Allocation for SWIPT in Multi-Group Multicast OFDM Systems. IEEE Commun Lett. 2017;21(11):2508–2511.

262 Hoang TM, El Shafie A, Duong TQ, Tuan HD, Marshall A. Security in MIMO-OFDM SWIPT Networks. In: IEEE 29th PIMRC; 2018. p. 1–6.

263 Khormuji MN, Popović BM, Perotti AG. Enabling SWIPT via OFDM-DC. In: IEEE WCNC; 2019. p. 1–6.

264 Buckley RF, Heath RW. System and Design for Selective OFDM SWIPT Transmission. IEEE Trans Green Commun Netw. 2021;5(1):335–347.

265 Ponnimbaduge Perera TD, Jayakody DNK, Sharma SK, Chatzinotas S, Li J. Simultaneous Wireless Information and Power Transfer (SWIPT): Recent Advances and Future Challenges. IEEE Commun Surveys Tuts. 2018;20(1):264–302.

266 Tang L, Zhang X, Zhu P, Wang X. Wireless Information and Energy Transfer in Fading Relay Channels. IEEE J Sel Areas Commun. 2016;34(12):3632–3645.

267 Zhou X, Li Q. Energy Efficiency for SWIPT in MIMO Two-Way Amplify-and-Forward Relay Networks. IEEE Trans Veh Technol. 2018;67(6):4910–4924.

268 Ding Z, Perlaza SM, Esnaola I, Poor HV. Power Allocation Strategies in Energy Harvesting Wireless Cooperative Networks. IEEE Trans Wireless Commun. 2014;13(2):846–860.

269 Mishra D, De S, Chiasserini CF. Joint Optimization Schemes for Cooperative Wireless Information and Power Transfer Over Rician Channels. IEEE Trans Commun. 2016;64(2):554–571.

270 Michalopoulos DS, Suraweera HA, Schober R. Relay Selection for Simultaneous Information Transmission and Wireless Energy Transfer: A Tradeoff Perspective. IEEE J Sel Areas Commun. 2015;33(8):1578–1594.

271 Wang D, Zhang R, Cheng X, Yang L, Chen C. Relay Selection in Full-Duplex Energy-Harvesting Two-Way Relay Networks. IEEE Trans Green Commun Netw. 2017;1(2):182–191.

272 Gupta A, Singh K, Sellathurai M. Time-Switching EH-Based Joint Relay Selection and Resource Allocation Algorithms for Multi-User Multi-Carrier AF Relay Networks. IEEE Trans Green Commun Netw. 2019;3(2):505–522.

273 Yuan Y, Xu Y, Yang Z, Xu P, Ding Z. Energy Efficiency Optimization in Full-Duplex User-Aided Cooperative SWIPT NOMA Systems. IEEE Trans Commun. 2019;67(8):5753–5767.

274 Shi L, Ye Y, Hu RQ, Zhang H. Energy Efficiency Maximization for SWIPT Enabled Two-Way DF Relaying. IEEE Signal Process Lett. 2019;26(5):755–759.

275 Li Q, Yang L. Robust Optimization for Energy Efficiency in MIMO Two-Way Relay Networks With SWIPT. IEEE Syst J. 2020;14(1):196–207.

276 Guo S, Zhou X, Zhou X. Energy-Efficient Resource Allocation in SWIPT Cooperative Wireless Networks. IEEE Syst J. 2020;14(3):4131–4142.

277 Zheng G, Krikidis I, Masouros C, Timotheou S, Toumpakaris DA, Ding Z. Rethinking the role of interference in wireless networks. IEEE Commun Mag. 2014;52(11):152–158.

278 Krikidis I. Simultaneous Information and Energy Transfer in Large-Scale Networks with/without Relaying. IEEE Trans Commun. 2014;62(3):900–912.

279 Alodeh M, Chatzinotas S, Ottersten B. Constructive Multiuser Interference in Symbol Level Precoding for the MISO Downlink Channel. IEEE Trans Signal Process. 2015;63(9):2239–2252.

280 Zhao N, Yu FR, Leung VCM. Opportunistic communications in interference alignment networks with wireless power transfer. IEEE Wireless Commun. 2015;22(1):88–95.

281 Khan TA, Alkhateeb A, Heath RW. Millimeter Wave Energy Harvesting. IEEE Trans Wireless Commun. 2016;15(9):6048–6062.

282 Alodeh M, Chatzinotas S, Ottersten B. Energy-Efficient Symbol-Level Precoding in Multiuser MISO Based on Relaxed Detection Region. IEEE Trans Wireless Commun. 2016;15(5):3755–3767.

283 Timotheou S, Zheng G, Masouros C, Krikidis I. Exploiting Constructive Interference for Simultaneous Wireless Information and Power Transfer in Multiuser Downlink Systems. IEEE J Sel Areas Commun. 2016;34(5):1772–1784.

284 Shojaeifard A, Wong KK, Di Renzo M, Zheng G, Hamdi KA, Tang J. Self-Interference in Full-Duplex Multi-User MIMO Channels. IEEE Commun Lett. 2017;21(4):841–844.

285 Varasteh M, Piovano E, Clerckx B. A Learning Approach to Wireless Information and Power Transfer Signal and System Design. In: IEEE ICASSP; 2019. p. 4534–4538.

286 Varasteh M, Hoydis J, Clerckx B. Learning Modulation Design for SWIPT with Nonlinear Energy Harvester: Large and Small Signal Power Regimes. In: IEEE 20th SPAWC; 2019. p. 1–5.

287 Varasteh M, Hoydis J, Clerckx B. Learning to Communicate and Energize: Modulation, Coding, and Multiple Access Designs for Wireless Information-Power Transmission. IEEE Trans Commun. 2020;68(11):6822–6839.

288 Willard J, Jia X, Xu S, Steinbach M, Kumar V. Integrating physics-based modeling with machine learning: A survey. arXiv preprint arXiv:200304919. 2020.

289 Durisi G, Koch T, Popovski P. Toward Massive, Ultrareliable, and Low-Latency Wireless Communication With Short Packets. Proc IEEE. 2016 sep;104(9):1711–1726.

290 Durisi G, Liva G, Polyanskiy Y. Short-packet transmission. Inf Theor Perspect 5G Syst Beyond. 2021. Available from: https://research.chalmers.se/en/publication/521844.

291 Durisi G, Koch T, Ostman J, Polyanskiy Y, Yang W. Short-Packet Communications Over Multiple-Antenna Rayleigh-Fading Channels. IEEE Trans Commun. 2016 feb;64(2):618–629.

292 Yang W, Caire G, Durisi G, Polyanskiy Y. Optimum Power Control at Finite Blocklength. IEEE Trans Inf Theory. 2015 sep;61(9):4598–4615.

293 Ostman J, Lancho A, Durisi G, Sanguinetti L. URLLC with Massive MIMO: Analysis and Design at Finite Blocklength. IEEE Trans Wirel Commun. 2021 sep:1–1.

294 Novosyolov A. The sum of dependent normal variables may be not normal. Institute of Computational Modelling, Siberian Branch of the Russian Academy of Sciences, Krasnoyarsk, Russia. 2006.

295 Chen X. Antenna correlation and its impact on multi-antenna system. PIER. 2015;62:241–253.

296 Kailath T. The Divergence and Bhattacharyya Distance Measures in Signal Selection. IEEE Trans Commun Technol. 1967 February;15(1):52–60.

297 Soman K, Diwakar S, Ajay V. Data mining: theory and practice [with CD]. PHI Learning Pvt. Ltd.; 2006.

298 De Amorim RC, Hennig C. Recovering the number of clusters in data sets with noise features using feature rescaling factors. Inf Sci. 2015;324:126–145.

299 Kaufman L, Rousseeuw PJ. Finding groups in data: an introduction to cluster analysis. vol. 344. John Wiley & Sons; 2009.

300 Al Shalabi L, Shaaban Z, Kasasbeh B. Data mining: A preprocessing engine. J Comput Sci. 2006;2(9):735–739.

301 Agarwal RP, Meehan M, O'Regan D. Fixed point theory and applications. vol. 141. Cambridge university press; 2001.

Index

Wireless RF Energy Transfer in the Massive IoT Era: Towards Sustainable Zero-energy Networks.
First Edition. Onel Alcaraz López and Hirley Alves.
© 2022 by The Institute of Electrical and Electronics Engineers, Inc. Published 2022 by John
Wiley & Sons, Inc. Companion website: www.wiley.com/go/Alves/WirelessEnergyTransfer

Printed and bound by CPI Group (UK) Ltd, Croydon, CR0 4YY